Genomes and Databases on the Internet

A Practical Guide to Functions and Applications

Paul Rangel and Jeremy Giovannetti

University California
Berkley
USA

Copyright © 2002
Horizon Scientific Press
P.O. Box 1
Wymondham
Norfolk NR18 0EH
England

www.horizonpress.com

Distributed exclusively in the United States, its dependent territories, Canada, Mexico, Central and South America, and the Caribbean by Springer-Verlag New York Inc, 175 Fifth Avenue, New York, USA, by arrangement with BIOS Scientific Publishers Ltd, 9 Newtec Place, Magdalen Road, Oxford OX4 1RE, UK.

Distributed exclusively in the rest of the world by BIOS Scientific Publishers Ltd, 9 Newtec Place, Magdalen Road, Oxford OX4 1RE, UK.

British Library Cataloguing-in-Publication Data

A catalogue record for this book is available from the British Library

ISBN: 1-898486-31-X

Printed and bound in Great Britain
by Antony Rowe Ltd, Chippenham, Wiltshire

Contents

Part II Genomic Research

Part III Online Analysis Tools

Part IV General Resources

Books of Related Interest

For further information on these books contact:

Horizon Scientific Press
P.O. Box 1, Wymondham
Norfolk
NR18 0EH England

Tel: +44(0)1953-601106
Fax: +44(0)1953-603068
Email: mail@horizonpress.com
Internet: www.horizonpress.com

**Our Web site has details of all our books including full chapter abstracts, book
reviews, and ordering information:
www.horizonpress.com**

Acknowledgements

We thank Horizon Scientific Press for putting together a project that caters to the people who need it most. We thank those of you who offered a hand in this project and those of you whose hand was led into the project. Thank you Dov Bock, Rod Kumimoto, Mark Leibman, Krista Lunde, Richard Mains, Sanjeev Pillai, Emily Queen, and Keri Schwartz. Special thanks go to Scarlett Chidgey, Betsy Cory, and Erica Tyler for editorial assistance.

Introduction

Genomes and Databases takes a practical look at what the web has to offer to the ever-expanding field of molecular biology. *Genomes* refers to the sequencing efforts of organisms and *Databases* refers to the storage of associated data. Nucleotide sequence data is the overwhelming result of genomics research and leads off our dissection of the web. The resources generated in response to nucleotide sequence data make up the rest of the book. They range from databases to analytical tools, protein sequences to metabolic pathways, and literature searches to molecular biology portals. We cover those resources that are freely available over the Internet.

Our perspective is that of someone who is aware of a few web offerings, has probably used a database or two, but is tepid in their approach to the expanse of the web. It is difficult enough to keep up with lab work and current literature, let alone find the time to scour the Internet for useful resources. We acquaint you with resources that can aid your research.

Our analysis, centered on genomes, databases, and applications to analyze their data, led to a four-part distribution:

Part I—Molecule Databases
Part II—Genomic Research
Part III—Online Analysis Tools
Part IV—General Resources

Part I, Molecule Databases covers all databases pertaining to DNA, RNA, and proteins. The primary nucleotide sequence databases, GenBank, EMBL, and DDBJ, are the most familiar ones, but we also

From: *Genomes and Databases on the Internet: A Practical Guide to Functions and Applications*
ISBN 1-898486-31-X © 2002 Horizon Scientific Press, Wymondham, UK.

look at protein sequence databases, PIR-PSD and SWISS-PROT, secondary nucleotide sequence databases, protein classification databases, structure databases, and databases focused on the function of genes.

Part II, Genomic Research takes a close look at the databases that present the results of genome sequencing projects. The human genome is just one of hundreds of genomes either fully sequenced or in the pipeline. Along with human genome databases, we analyze databases dedicated to the genomes of plants, fungi, vertebrates, invertebrates, insects, organelles, bacteria, and viruses. We also review genome centers that handle a multitude of genomes and provide comparative genome analyses.

Part III, Online Analysis Tools explains various software applications that can be used in genomic and proteomic research. Tool functions may relate to nucleotide and protein sequence alignments or phylogenetic tree development or even the modeling of three-dimensional protein structures. There are a number of tools with a breadth of uses that we organize and present as they relate to specific tasks.

Part IV, General Resources contains a brief chapter that examines web resources beyond molecules. The "databases" in this chapter contain web links, literature references, and lab supplies. We also denote some of the online publications of genome research.

Because the book is based on the Internet, we begin the tour with a chapter explaining some of the things that make the web tick in *Web Basics*.

Part I

Molecule Databases

Chapter 1

Web Basics

Contents

Abstract

The exponential growth of the Internet and nucleotide sequence data has changed the face of molecular biology. Wet labs in many research facilities are welcoming computational biologists and exchanging the sterile hood space for computer server racks. With this new technology comes new terminology and more room for confusion. This chapter

From: *Genomes and Databases on the Internet: A Practical Guide to Functions and Applications*
ISBN 1-898486-31-X © 2002 Horizon Scientific Press, Wymondham, UK.

is a quick review of the fundamental Internet technologies that the book will build upon when describing the tools available for Internet research. One aspect of the technology, the client – server relationship, has defined the way that personal computers interact with remote computers around the world. Other technologies such as the Hyper Text Transfer Protocol (HTTP) and File Transfer Protocol (FTP) define the standards for file sharing between foreign computers. Combined, these technologies make the Internet and computational molecular biology possible.

Introduction

It is no accident the genomic revolution was tightly coupled with the prolific growth of the Internet. Without a means to distribute data, the genome sequencing projects would have faced many more problems than accurate sequence annotation. It is because of this tight relationship that we review some of the technologies that support our ability to access data from around the world.

Client / Server Architecture

The web was originally built around the assumption of a client/server relationship between two computers. The client requests information or launches applications on the server and the server responds with the appropriate data. This is most often seen in the web server-web client interaction where web pages are requested from remote computers and a web page of text and images is returned. Typically a genome server is a high-powered computer with web-enabled applications such as a database of sequence records. Other types of client-server applications include FTP applications and database applications. In contrast, the peer-to-peer relationship, when computers openly share files back and forth, is less popular among genome database users. A widespread application of this technology is for music and image file sharing.

Protocols

Protocols used in the lab ensure that the outcome of our work is the same as the authors. This same strategy is used over the Internet when communicating with other computers. Both ends of the connection follow protocols to ensure that the data received is the same as the data sent. A more concise definition from *Webopedia* (http://webopedia.internet.com) is "An agreed upon format for transmitting data between two devices." Protocols come in many varieties and are constantly adapting to new technologies.

HTTP

If you have used a web browser, Internet Explorer or Netscape, you have used the Hyper Text Transfer Protocol (HTTP). This is the protocol that allows you to download web pages from another computer with a web page server installed. HTTP is translated by your browser when received from another computer and rendered as an interactive compilation of text and graphics.

FTP

The File Transfer Protocol (FTP) facilitates the transfer of files across the Internet. There are many ways to use FTP. The latest versions of web browsers will let you view FTP file systems (ftp://…) from the web browser window enabling direct downloads to your hard drive. There are also FTP clients such as FETCH and WUFTP that can be downloaded and installed on your computer for direct file access. Many of the online resources we discuss throughout the book can be downloaded to your computer via FTP.

CGI

The Common Gateway Interface (CGI) describes the specifications to transfer information from a web server to an application. This interface provides a means of executing programs on a remote computer in a protected environment.

Technologies

Web Standards

Web standards are somewhat of a misnomer. As the web has grown, various companies and software development groups have implemented data sharing standards. Unfortunately, few have risen to the challenge to agree on any one standard. This has left the details of the Internet in tangles. Enter the World Wide Web Consortium (W3C; http://www/w3.org), an independent group of web leaders who propose standards for data on the web. A visit to the W3C site explains many of technicalities of web standards and is a great way to learn about new technologies being adopted on the web.

The Web is Stateless

Each request to a web server is unique meaning that the information about the client is not held in the web server. This is important when trying to use web-based applications that require multiple stages of user input. If an application does not implement a means to record a transaction, the state of the user interaction is lost and there is no way to continue an interrupted or previous request. Technologies such as cookies, Javascript and Java servlets help with this problem, but they require client side participation.

Javascript vs. Java

One of the new developments from molecular biology database projects has been an interactive graphical depiction of a genome. This window into the sequence data provides active links and buttons for browsing segments of chromosomes and in some cases performing alignment analyses. These applications have made use of two similar but unrelated technologies, Javascript and Java.

Javascript is a programming language used to control the browser window of your computer. It can change the look of buttons when they are clicked and launch new browser windows when needed, but it is confined to the browser window. It cannot establish network connections with other computers or create dynamic images. Many genome browsers use Javascript to control the content of a page or launch additional windows with related information.

Java is a programming language developed by Sun Microsystems. It runs on most operating systems and is becoming more popular among databases. Java programs, known as Java Applets, are launched through web pages and have their own control mechanism much like a regular application. These can be in the form of non-HTML buttons or images that may be clicked or dragged to perform functions not possible in a normal HTML window. Because Applets are complete programs, they are often larger than a typical web page taking more time to download and additional processing power to run.

Chapter 2

Primary Nucleotide Sequence Databases

Contents

From: *Genomes and Databases on the Internet: A Practical Guide to Functions and Applications*
ISBN 1-898486-31-X © 2002 Horizon Scientific Press, Wymondham, UK.

Abstract

Primary sequence databases are the cornerstone of bioinformatics research. Databases such as GenBank and EMBL accept genome data from sequencing projects around the world and make it available for researchers via the World Wide Web. The underlying organization of these databases has shaped the way computer-based molecular biology research is conducted both at these facilities and in related secondary databases. Understanding primary nucleotide sequence databases is key in understanding molecular biology on the web.

Introduction

If the laboratory is the foundation of experimental biology, primary nucleotide sequence databases are the foundation of computational biology. These databases supply raw genetic data—nucleotide sequence—and a variety of resources to extract information from it. Simple questions relating to subjects such as presence or absence of similar sequences, amount of genetic data available for an organism, and relation to genes can be answered through the primary nucleotide sequence databases. First and foremost, these databases provide a comprehensive resource for publicly available nucleotide sequence data. These databases, of which at least one is probably familiar to the reader, have been fully operational since the 1980s.

- GenBank at NCBI (National Center for Biotechnology Informatics)
- EMBL (European Molecular Biology Laboratories) Nucleotide Database at the EBI (European Bioinformatics Institute)
- DDBJ (DNA Data Bank of Japan) at the CIB (Center for Information Biology)

The formation of a centralized biological repository for molecular biology data was a natural process. The rapid advances in DNA sequencing technology, the establishment of a world data network, and the development of inexpensive high-powered personal computers united scientific research and catalyzed the "omic" revolution. Genomics, at the forefront of the revolution, flooded the primary genetic databanks with an incredible amount of sequence data, which via the World Wide Web, was distributed to the public. In 1988, representatives of the three centers, now known as the International Nucleotide Sequencing Database Consortium (INSDC), met and formalized a common format for describing the nucleotide data in their databases. This decision facilitated the distribution of genetic data around the world by giving researchers a universal language by which they could share it.

This chapter describes some of the tools used to distribute genetic data and discusses the capabilities of the three nucleotide sequence centers that are the backbone of molecular biology resources on the Internet.

International Collaboration

The triumvirate responsible for all genetic data, GenBank, EMBL, and DDBJ, is an international consortium (Figure 2.1). It is due to diligent communication and cooperation that so much data is clearly, comprehensively, and accurately made available to researchers at their personal computers.

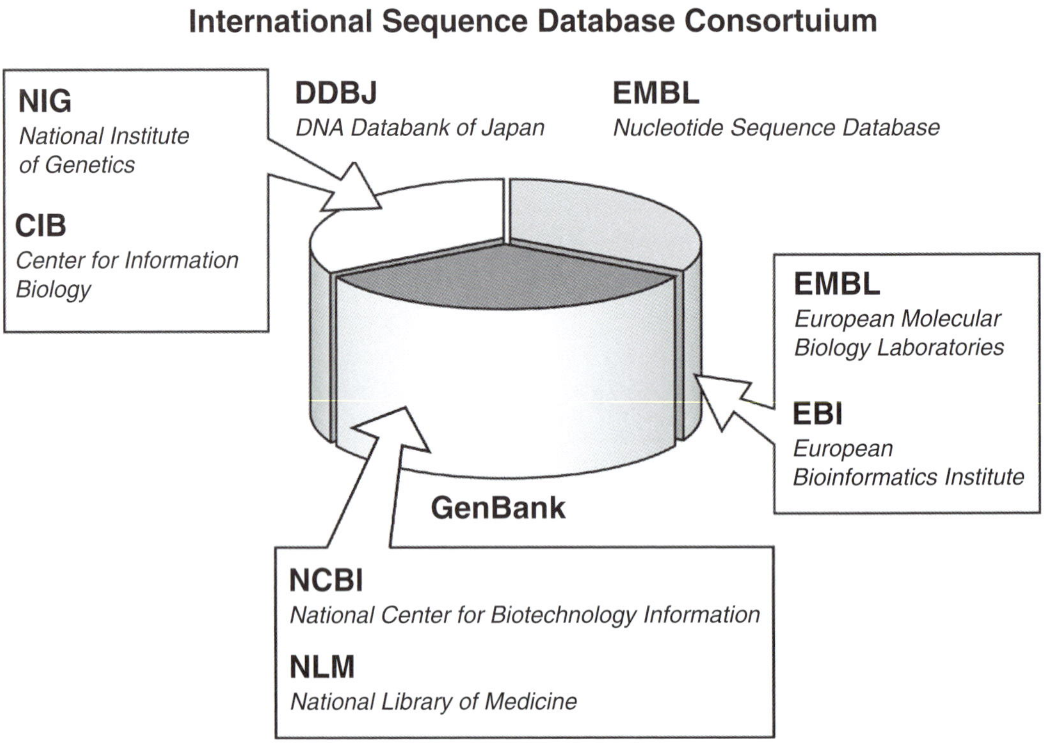

International Sequence Database Consortuium

NIG
National Institute of Genetics

CIB
Center for Information Biology

DDBJ
DNA Databank of Japan

EMBL
Nucleotide Sequence Database

EMBL
European Molecular Biology Laboratories

EBI
European Bioinformatics Institute

GenBank

NCBI
National Center for Biotechnology Information

NLM
National Library of Medicine

Figure 2.1. The International Sequence Database Consortium (INSDC) participants. All nucleotide sequence data is shared between GenBank, EMBL, and DDBJ.

The International Nucleotide Sequence Database Consortium (INSDC), the united name of the trio, collectively obtains, processes, and publishes data for public use. Labs worldwide generate sequence data that is eventually submitted to the INSDC. The consortium has set up a way to share data efficiently so that each database is consistently supporting the same data set.

The collaboration among the three allows each to develop their own web interfaces that uniquely represent the data as well as develop varying resources to analyze it after the sequences have been shared.

It is worth noting the roles of the INSDC as both primary and secondary sequence databases. The centers serve as the world's primary or archival repository for genetic data while employing scientists to curate the data for secondary databases such as the Cancer Chromosome Aberration Project and the Conserved Domain Database.

Data Sharing

Collaboration is a monumental task requiring enough foresight to allow for the exponential growth of sequence data while continuously updating (sequences are shared each day) three diverse data centers. A great deal of the success the INSDC has experienced can be attributed to its ability to decide upon and adhere to a common data format. Knowledge of the terminology and layout of the sequence files is useful because it provides an understanding of the source formats for genetic data encountered in many other databases on the Internet.

ASN.1

As is seen Figure 2.2, the task of handling all of the data has become significantly more difficult over the past 10 years. To complicate matters, this information must be shared across heterogeneous computer systems. As we have seen in Chapter 1, Web Basics, the best way to accomplish this is with a common language such as the ASN.1 (Abstract Syntax Notation One), derived by the collaboration. It is similar to the HTML standard because it is a set of rules followed by a computer that describes information shared by two unknown users. This genomic syntax provides the data infrastructure to facilitate storage and retrieval of sequence, structure, literature, and genomic data from the distributed databases. After interpretation by a program, the output is formatted into a human readable page such as the GenBank output of a BLAST search (Basic Local Alignment Search Tool). Again, this compares to source HTML code entering your web browser as raw text and being displayed as a formatted web page.

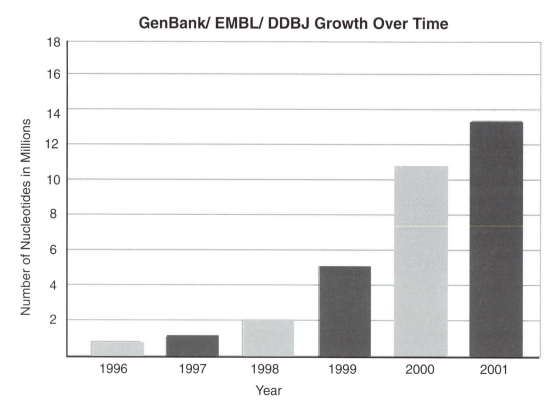

http://www.ddbj.nig.ac.jp/images/ddbjnew/DBGrowth-e.gif

Figure 2.2. Estimated growth of the nucleotide sequence data at the major sequencing facilities

Features Table

The ASN.1 describes the format of the data as it is transmitted. The format was defined based on the type of data being shared. An agreed upon table of features that describes the DNA and protein sequence annotations (coding sequence, organism, tissue type, gene, etc.) is used by the collaboration to share information while maintaining an accurate account of the individual data elements such as the Coding Sequence (CDS) and Organism (ORGANISM). An agreed upon features table allows a researcher to use either a GenBank or EMBL format with the assurance that the individual elements are consistent. This standard format for the table allows each of the data centers to create a unique format for their computing purposes while maintaining the structure of the data for distribution among the three centers.

Taxonomy Database

The Taxonomy Database project has been adopted by the sequence databases primarily to have a consistent set of names for the 74,000 organisms one might have sequence for. Submitted data includes the organism the sequence came from, which must match with a name found in the Taxonomy Database. This makes sifting through files in a search an accurate process. The project also assists with the translation of the nucleotide sequences into proteins.

Display to the Public

Above are the technologies used to facilitate the daily distributions of the sequence data between the three primary sequence repositories. Much like the raw HTML of a web page, the ASN.1 file remains invisible to the users. The data from these files is parsed and formatted at the sequence repository for display to the public. The display of a sequence and the data that describes it is the GenBank/DDBJ flat file (almost identical) and the EMBL flat file. This file type is what is encountered when searching the databases for sequences of interest, but there are variations as to how much data a user wants to see.

Sometimes the most useful display of the data is simply the sequence and a short identifier. This is useful for applications that perform a similarity search against a database. FASTA is the most common of the sequence-only flat file displays to be used in analysis and search programs such as BLAST and Entrez. It is worth noting that each of the repositories utilizes the data in their own way and each provides the results of a search on the data in a unique format known as the database flat file. The human readable format of the GenBank/DDBJ flat file and the EMBL flat file is the representation of the data as researchers see it over the Internet. (Remember that the formats are

```
>gi|14724116|ref|XM_052368.1| Homo sapiens alcohol dehydrogenase 1A
GAAGACAGAATCAACATGAGCACAGCAGGAAAAGTAATCAAATGCAAAGCAGCTGTGCTATGGGAGTTAA
AGAAACCCTTTTCCATTGAGGAGGTGGAGGTTGCACCTCCTAAGGCCCATGAAGTTCGTATTAAGATGGT
GGCTGTAGGAATCTGTGGCACAGATGACCACGTGGTTAGTGGTACCATGGTGACCCCACTTCCTGTGATT
TTAGGCCATGAGGCAGCCGGCATCGTGGAGAGTGTTGGAGAAGGGGTGACTACAGTCAAACCAGGTGATA
AAGTCATCCCACTCGCTATTCCTCAGTGTGGAAAATGCAGAATTTGTAAAAACCCGGAGAGCAACTACTG
CTTGAAAAACGATGTAAGCAATCCTCAGGGGACCCTGCAGGATGGCACCAGCAGGTTCACCTGCAGGAGG
AAGCCCATCCACCACTTCCTTGGCATCAGCACCTTCTCACAGTACACAGTGGTGGATGAAAATGCAGTAG
CCAAAATTGATGCAGCCTCGCCTCTAGAGAAAGTCTGTCTCATTGGCTGTGGATTTTCAACTGGTTATGG
GTCTGCAGTCAATGTTGCCAAGGTCACCCCAGGCTCTACCTGTGCTGTGTTTGGCCTGGGAGGGGTCGGC
CTATCTGCTATTATGGGCTGTAAAGCAGCTGGGGCAGCCAGAATCATTGCGGTGGACATCAACAAGGACA
AATTTGCAAAGGCCAAAGAGTTGGGTGCCACTGAATGCATCAACCCTCAAGACTACAAGAAACCCATCCA
GGAGGTGCTAAAGGAAATGACTGATGGAGGTGTGGATTTTTCATTTGAAGTCATCGGTCGGCTTGACACC
ATGATGGCTTCCCTGTTATGTTGTCATGAGGCATGTGGCACAAGTGTCATCGTAGGGGTACCTCCTGATT
CCCAAAACCTCTCAATGAACCCTATGCTGCTACTGACTGGACGTACCTGGAAGGGAGCTATTCTTGGTGG
CTTTAAAAGTAAAGAATGTGTCCCAAAACTTGTGGCTGATTTTATGGCTAAGAAGTTTTCATTGGATGCA
TTAATAACCCATGTTTTACCTTTTGAAAAAATAAATGAAGGATTTGACCTGCTTCACTCTGGGAAAAGTA
TCCGTACCATTCTGATGTTTTGAGACAATACAGATGTTTTCCCTTGTGGCAGTCTTCAGCCTCCTCTACC
CTACATGATCTGGAGCAACAGCTGGGAAATATCATTAATTCTGCTCATCACAGATTTTATCAATAAATTA
CATTTGGGGGCTTTCCAAAGAAATGGAAATTGATGTAAAATTATTTTTCAAGCAAATGTTTAAAATCCAA
ATGAGAACTAAATAAAGTGTTGAACATCAGCTGGGGAATTGAAGCCAATAAACCTTCCTTCTTAACCATT
```

Figure 2.3. An example of the FASTA format. Note the top line begins with ">" and serves to identify the nucleotide sequence below.

just human readable versions of the ASN.1 format used by the databases to store data). This flat file format is also the mechanism of data transfer to secondary databases. An understanding of flat file structure assists in the interpretation of many of the results returned by web based research and gives a good general understanding of the database data model.

File Formats

File formats are the language used by the data centers to communicate to the world. Secondary databases, public applications, and commercial applications have been built around the data formats used at these facilities. The structure of two of the most common formats, FASTA and GenBank, illustrate the different methods of data display and how an understanding of file formats can assist in online research.

FASTA is the simplest of the file formats and generally is the default format used in web-based applications. The first line of the file begins

HEADER
```
LOCUS       XM_052368     1400 bp    mRNA            PRI       16-JUL-2001
DEFINITION  Homo sapiens alcohol dehydrogenase 1A (class I), alpha polypeptide
            (ADH1A), mRNA.
ACCESSION   XM_052368
VERSION     XM_052368.1  GI:14724116
KEYWORDS    .
SOURCE      human.
  ORGANISM  Homo sapiens
            Eukaryota; Metazoa; Chordata; Craniata; Vertebrata; Euteleostomi;
            Mammalia; Eutheria; Primates; Catarrhini; Hominidae; Homo.
REFERENCE   1  (bases 1 to 1400)
  AUTHORS   NCBI Annotation Project.
  TITLE     Direct Submission
  JOURNAL   Submitted (12-JUL-2001) National Center for Biotechnology
            Information, NIH, Bethesda, MD 20894, USA
```

BODY
```
FEATURES             Location/Qualifiers
     source          1..1400
                     /organism="Homo sapiens"
                     /db_xref="taxon:9606"
                     /chromosome="4"
     gene            1..1400
                     /gene="ADH1A"
                     /note="ADH1"
                     /db_xref="LocusID:124"
                     /db_xref="MIM:103700"
     CDS             16..1143
                     /gene="ADH1A"
                     /codon_start=1
                     /product="class I alcohol dehydrogenase, alpha subunit"
                     /protein_id="XP_052368.1"
                     /db_xref="GI:14724117"
                     /translation="MSTAGKVIKCKAAVLWELKKPFSIEEVEVAPPKAHEVRIKMVAV
                     GICGTDDHVVSGTMVTPLPVILGHEAAGIVESVGEGVTTVKPGDKVIPLAIPQCGKCR
                     ICKNPESNYCLKNDVSNPQGTLQDGTSRFTCRRKPIHHFLGISTFSQYTVVDENAVAK
                     IDAASPLEKVCLIGCGFSTGYGSAVNVAKVTPGSTCAVFGLGGVGLSAIMGCKAAGAA
                     RIIAVDINKDKFAKAKELGATECINPQDYKKPIQEVLKEMTDGGVDFSFEVIGRLDTM
                     MASLLCCHEACGTSVIVGVPPDSQNLSMNPMLLLTGRTWKGAILGGFKSKECVPKLVA
                     DFMAKKFSLDALITHVLPFEKINEGFDLLHSGKSIRTILMF"
```

FOOTER
```
BASE COUNT      400 a     294 c     339 g     367 t
ORIGIN
        1 gaagacagaa tcaacatgag cacagcagga aaagtaatca aatgcaaagc agctgtgcta
       61 tgggagttaa agaaaccctt ttccattgag gaggtggagg ttgcacctcc taaggcccat
      121 gaagttcgta ttaagatggt ggctgtagga atctgtggca cagatgacca cgtggttagt
      181 ggtaccatgg tgaccccact tcctgtgatt ttaggccatg aggcagccgg catcgtggag
      241 agtgttggag aagggtgac tacagtcaaa ccaggtgata aagtcatccc actcgctatt
      301 cctcagtgtg gaaaatgcag aatttgtaaa aaccggaga gcaactactg cttgaaaaac
      361 gatgtaagca atcctcaggg gaccctgcag gatggcacca gcaggttcac ctgcaggagg
      421 aagcccatcc accacttcct tggcatcagc accttctcac agtacacagt ggtggatgaa
      481 aatgcagtag ccaaaattga tgcagcctcg cctctagaga aagtctgtct cattggctgt
      541 ggattttcaa ctggttatgg gtctgcagtc aatgttgcca aggtcacccc aggctctacc
      601 tgtgctgtgt ttggcctggg aggggtcggc ctatctgcta ttatgggctg taaagcagct
      661 gggcagcca gaatcattgc ggtggacatc aacaaggaca aatttgcaaa ggccaaagag
      721 ttgggtgcca ctgaatgcat caaccctcaa gactacaaga aacccatcca ggaggtgcta
      781 aaggaaatga ctgatgagg tgtggatttt tcatttgaag tcatcggtcg gcttgacacc
      841 atgatggctt ccctgttatg ttgtcatgag gcatgtggca caagtgtcat cgtaggggta
      901 cctcctgatt cccaaaacct ctcaatgaac cctatgctgc tactgactgg acgtacctgg
      961 aagggagcta ttcttggtgg ctttaaaagt aaagaatgtg tcccaaaact tgtggctgat
     1021 tttatggcta agaagttttc attggatgca ttaataaccc atgttttacc ttttgaaaaa
     1081 ataaatgaag gatttgacct gcttcactct gggaaaagta tccgtaccat tctgatgttt
     1141 tgagacaata cagatgtttt cccttgtggc agtcttcagc ctcctctacc ctacatgatc
     1201 tggagcaaca gctgggaaat atcattaatt ctgctcatca cagattttat caataaatta
     1261 catttggggg cttttccaaag aaatgtgaaat tgatgtaaaa ttatttttca agcaaatgtt
     1321 taaaatccaa atgagaacta aataagtgt tgaacatcag ctggggaatt gaagccaata
     1381 aaccttcctt cttaaccatt
```

Figure 2.4. An example of the GenBank Flat File (GBFF) format. It is divided into a header, body, and footer.

with the "greater than" symbol (>) followed by a description of the sequence (see Figure 2.3). The actual sequence data starts on the second line. Programs such as BLAST accept the FASTA formats and can use the first line of the file when identifying the results of a search. This feature can help when processing large amounts of sequence where many unique results are returned.

The GenBank Flat File

Because the file formats of the three databases are similar, only the GenBank Flat File (GBFF) will be described (see Figure 2.4). Unique or helpful features of the EBML and DDBJ formats are mentioned where appropriate.

GBFF Header

The header of the GenBank file is meant to guide the researcher by describing the non-molecular features of the entry. This includes information about the origins of the sequence, the identification of the source organism, and the unique identifiers associated with the record.

Locus

The beginning of every entry starts with the locus tag, which currently consists of an entry's accession number followed by the sequence length and molecule type. The three-letter division tag is a GenBank specific tag that aids in the storage and retrieval of the entries. Divisions are useful in performing queries on subsets of the GenBank database. The date at the end of *LOCUS* represents the last modified date for the record or, if there have not been modifications, the entry date of the sequence. NCBI must be contacted directly to confirm the original entry date of the sequence.

Definition

The definition is a brief description of the record. The information on this line is also used as the header of all the FASTA formats and BLAST result searches. The line begins with the genus and species names of the source organism, followed by the gene product name and other notes. A completion qualifier is added if the sequencing of the gene has been completed. The EMBL equivalent of the *DEFINITION* line is the *DE* (description) line.

Accession

The accession number is the persistent tag for each record in all of the databases. It is seven characters long with either one letter followed by six digits or two letters followed by five digits. There are secondary accession numbers that indicate a change of the original entry data.

Version

Ever-shifting gene annotations have created the need for a system to keep track of the changes in gene sequence. The version number is attached to the end of the accession number, separated by a period. A GI number (GenInfo Identifier) is also added to a record if it has been changed in any way.

Source organism

The corresponding EMBL line is also called *ORGANISM*. The databases utilize the NCBI Taxonomy Database to correctly describe the lineage of the organism for each record.

Reference

Each of the records in the GenBank database has at least one associated reference or citation that describes the laboratory or publication of origin.

GBFF Body

Features

The most informative area of the GBFF is the Features Table. As described in the introductory section, the features in the table are used by all of members of the INSDC. One or more qualifiers may accompany each of the feature elements, which allows for a further description of it. The feature is aligned to the left side of the document with the corresponding sequence located directly across; the qualifiers are listed directly below separated by a forward slash (/qualifier = "qualifier text"). To assist with annotations, data contributors are asked to provide as much of the feature information as possible before submitting the entry into the database. WebFeat at EBI and the Sequin Help documentation at GenBank, both listed in Table 2.1, assist with the annotation process by outlining the features and qualifiers needed for a successful database entry.

Three important Features of the Features Table are *source*, *gene*, and *CDS* (Coding Sequence). *Source* is the only required feature of the table and one of the few features with a mandatory qualifier. *Source* depicts the biological source of the sequence beginning with the required organism qualifier. Some of the optional qualifiers include cell_line, plasmid, sex, strain, tissue_type etc. *gene* is described as a range in the nucleotide sequence that has been identified as a gene, for which there is a corresponding name. The gene qualifiers include allele, map, product and partial. *CDS* describes an area of a sequence that has been translated from a nucleotide chain to a sequence of amino acids.

GBFF Footer

The end of the document displays a raw summary of the sequence data beginning with the *BASE COUNT,* or the number of each of the bases in the nucleotide sequence. The sequence follows in a format very similar to that of FASTA. The end of the GBFF is denoted by two forward slashes on the last line of the page.

Database Resources

We have introduced the basic tenets of the three international sequence databases that include the technologies upon which they run and some of the common file formats that display their sequence data. While the databases are the primary focus of the INSDC, the research facilities of the centers have pioneered the development of sequence data mining tools that have become an integral part of computational biology over the web. These database resources have created new levels of database integration and searching capabilities. Examples of these are the Sequence Retrieval System (SRS), developed at EMBL, and Entrez, developed at NCBI (see Chapter 10 for more information on SRS and Entrez). Secondary databases such as Ensembl, the open source human genome server, and OMIM (On the Mendilian Inheritance in Man) are active projects and are additional examples of the online research initiated at these facilities.

Below are some of the auxiliary facets of the INSDC research centers.

Data Submission

Collection of new sequence data is essential for the development of an online database resource. The INSDC databases collect information from sources around the world and have each developed systems for researches to enter data from remote computers. Data submission systems involving the web, email, or stand-alone software are the most frequently used by individual researchers or small labs. Larger

Table 2.1. Primary Nucleotide Sequence Databases

#	Site Name	URL
1.	GenBank	http://www.ncbi.nlm.nih.gov/Genbank/GenbankOverview.html
2.	EMBL – European Molecular Biology Laboratory	http://www.embl-heidelberg.de/
3.	DDBJ – DNA Data Bank of Japan	http://www.ddbj.nig.ac.jp/

Related Sites

#	Site Name	URL
4.	INSDC – International Nucleotide Sequencing Database Consortium	http://www.ncbi.nlm.nih.gov/collab/
5.	Features Table	http://www.ncbi.nlm.nih.gov/collab/FT/index.html
6.	WebFeat	http://www3.ebi.ac.uk/Services/WebFeat/
7.	Taxonomy Database	http://www.ncbi.nlm.nih.gov/Taxonomy/
8.	EMVEC	http://www2.ebi.ac.uk/blastall/vectors.html
9.	Webin	http://www.ebi.ac.uk/embl/Submission/webin.html
10.	BankIt	http://www.ncbi.nlm.nih.gov/BankIt/
11.	Email Data Submission	http://www.ncbi.nlm.nih.gov/Genbank/
12.	Sequin	http://www.ncbi.nlm.nih.gov/Sequin/index.html
13.	BLAST	http://www.ncbi.nlm.nih.gov/BLAST/
15.	SRS	http://srs6.ebi.ac.uk/
16.	Entrez	http://www.ncbi.nlm.nih.gov/Entrez/
17.	CLUSTAL	http://www.ebi.ac.uk/clustalw/
18.	EnsEMBL	http://www.ncbi.nlm.nih.gov/Entrez/
19.	OMIM	http://www.ncbi.nlm.nih.gov/entrez/query.fcgi?db=OMIM
20.	Human Genome Map Viewer	http://www.ncbi.nlm.nih.gov/entrez/query.fcgi?db=Genome
21.	Human Genome Studio	http://studio.nig.ac.jp/index.html

data submissions from genome sequencing centers, for example, have agreements to help facilitate sequence deposition.

All submitted data is manually curated by a team of administrators and biologists according to rules established by the INSDC. The curation process checks for redundant entries and data integrity as well as cross-references the entry to other sources of data among other duties.

An important part of the submission process is screening for junk DNA. The source of the junk DNA, for example, can be from the vector used to sequence the DNA fragment of interest. EBI combats these entries by screening the submission against the EMVEC vector database. NCBI has a similar vector screening method. These methods make up the front line of defense against sequence contaminants.

Web Submission: Webin and BankIt

The web is the easiest and most convenient method of sequence submission. Webin at EBI and BankIt at NCBI are two web-based applications for submitting sequences in bulk or individually.

E-mail

Email submission is conducted via an email. While this method is not the suggested method for regular data entry because of the lack of automation in the procedure, it is recommended for submitters without a reliable connection to the Internet. A sample of the form can be found at all the centers or requested via email.

Software: Sequin

NCBI has developed a stand-alone software submission system to assist with the sequence submission process. Sequin can be downloaded and installed on many operating systems including Macintosh, Windows, Linux, and Unix. This system is best used when submitting complex sequence information such as sequence alignments.

Sequin has uses as an analysis tool as well. There is a "network aware" version of sequin that can be used as a NCBI browser. Through its interface you can access and edit existing GenBank data, run BLAST searches, use the NBCI Taxonomy Database and scan for vector sequences.

Sequence Similarity Tools

Sequence similarity tools assist sequence analysis by providing a mechanism to filter or align sequence information. The database centers make available tools such as BLAST and CLUSTALW for these purposes. See Chapter 10, Genomics Tools, for more information on how these tools work.

Secondary Databases

Genome annotations, EST compilations, and gene clustering research create secondary databases of all varieties. The Human Genome Project has spawned a number of secondary databases such as ENSEMBL, Online Mendelian Inheritance in Man (OMIM), the Human Genome MapViewer, and the Human Genome Studio. See Table 2.1 for URLs.

Taxonomy Browsers

Taxonomy has served as an intuitive entry point into the three databases. It is interesting to note that all three of the centers have developed taxonomy tools in unique formats. For example, the NCBI Taxonomy Browser (for URL see Table 2.1) provides a user-friendly interface accessible via the tool bar at the top of the page. The output of the results is a taxonomic summary of the species in question. The summary data can subsequently be used to narrow BLAST searches by setting the database parameters to a subset of taxonomic data.

Further Reading

Baxevanis, A.D., and Ouellette, B.F.F. eds. 1998. Bioinformatics: A Practical Guide to the Analysis of Genes and Proteins. Wiley-Interscience, John Wiley and Sons, New York.

Stoesser, G., Baker, W., Van den Broek, A., Camon, E., Garcia-Pastor, M., Kanz, C. Kulikova, T., Lombard, V., Lopez, R., Parkinson, H., Redaschi, N., Sterk, P., Stoehr, P., and Tuli, M. 2001. The EMBL nucleotide sequence database. Nucleic Acids Res. 29: 17-21.

Tateno, Y., Miyazaki, S., Ota, M., Sugawara, H., and Gojobori, T. 2000. DNA Data Bank of Japan (DDBJ) in collaboration with mass sequencing teams. Nucleic Acids Res. 28: 24-26.

Wheeler, D.L., Church, D.M., Lash, A.E., Leipe, D.D., Madden, T.L., Pontius, J.U., Schuler, G.D., Schriml, L.M., Tatusova, T.A., Wagner, L., and Rapp, B.A. 2001. Database resources of the National Center for Biotechnology Information. Nucleic Acids Res. 29: 11-16.

Chapter 3

Primary Protein Sequence Databases

Contents

From: *Genomes and Databases on the Internet: A Practical Guide to Functions and Applications*
ISBN 1-898486-31-X © 2002 Horizon Scientific Press, Wymondham, UK.

Abstract

Primary protein sequence databases are to protein sequences what GenBank, EMBL, and DDBJ are to nucleotide sequences. They are the central location of protein sequence data submissions. PIR's Protein Sequence Database (PSD) and SWISS-PROT are the two main databases. They provide a variety of ways to access data and analysis tools once you have retrieved the sequence you were looking for. A detailed review of accessing data through PIR's *Selection List* is provided. Other databases that are mentioned are OWL, Entrez' Protein database, and Peptide/Protein Sequence Database (RPF/ SEQDB).

Introduction

The primary protein sequence databases represent a complete set of publicly available sequence data. These are analogous to the primary nucleic acid databases GenBank, EMBL, and DDBJ: they are the entry point of new protein sequences to the web community. The Protein Information Resource's Protein Sequence Database (PIR-PSD) and SWISS-PROT (plus TrEMBL) are the two significant primary protein sequence databases.

These databases are essential for a variety of reasons within the scope of molecular evolution and functional genomics. For instance, the study of proteins, the biologically active state of a gene, is important in terms of determining the function of a gene.

The complete set of protein sequences for an organism is considered a proteome (versus a genome for nucleotides), and proteomics (versus genomics) defines the study of an organism at the protein level. Proteomes are briefly mentioned below in the review of SWISS-PROT, as the SWISS-PROT group initiated the Human Proteomics Initiative (HPI) and the High quality Automated Microbial Annotation of Proteomes (HAMAP). The foundation of proteomics is the primary protein sequence database since they receive the sequences. The proteome databases extract sequences specific to one organism. The classification databases of Chapter 5 also get their sequences from these databases.

The primary protein sequence databases provide a wealth of data about individual sequences. They go as far as classifying proteins into families and describing their function and structure. These attributes as well as other questions concerning these databases are addressed in detail below.

Primary Protein Sequence Databases

There are two primary protein sequence databases. They do not collaborate and are not complementary, but each essentially provides access to all publicly available protein sequences and data characterizing those proteins. Between the two, there can be slight differences in the annotation data for the same protein because different people are working on them, but as a whole the two databases are quite similar. Both databases deserve attention and ultimately users will choose which one works best for them. Our reviews aim to get those that are interested acquainted with how these databases can aid research by explaining the resources they provide and how to use them.

PIR-Protein Sequence Database

The Protein Information Resource (PIR) is a community resource that provides protein databases and analysis tools to support protein-based molecular biology. The Protein Sequence Database (PSD) is one of the most comprehensive resources available on the web for molecular biologists and is the most significant service offered by PIR.

The first collection of protein sequences was the *Atlas of Protein Sequence and Structure* published in 1965 by the National Biomedical

Annotation

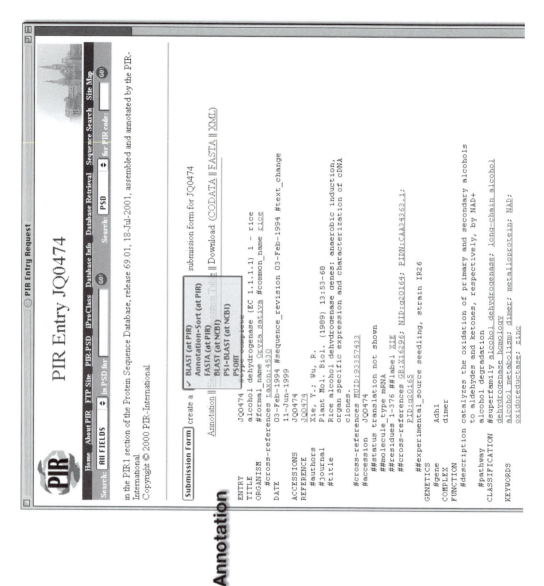

PIR Entry JQ0474

Home | About PIR | FTP Site | PIR-PSD | iProClass | Database Info | Database Retrieval | Sequence Search | Site Map

Search: All FIELDS ⬦ in PSD for [   ]   Search: PSD ⬦ for PIR code: [   ] GO!

in the PIR1 section of the Protein Sequence Database, release 69.01, 18-Jul-2001, assembled and annotated by the PIR-International

Copyright © 2000 PIR-International

Submission Form | create a _____ submission form for JQ0474

Annotation || Selection Button Table || Download (CODATA || FASTA || XML)

- BLAST (at PIR)
- Annotation-Sort (at PIR)
- FASTA (at PIR)
- BLAST (at NCBI)
- PSI-BLAST (at NCBI)
- PSORT

```
ENTRY            JQ0474          #type complete
TITLE            alcohol dehydrogenase (EC 1.1.1.1) 1 - rice
ORGANISM         #formal_name Oryza sativa #common_name rice
                 #cross-references taxon:4530
DATE             03-Feb-1994 #sequence_revision 03-Feb-1994 #text_change
                 11-Jun-1999
ACCESSIONS       JQ0474
REFERENCE        JQ0474
                 #authors Xie, Y.; Wu, R.
                 #journal Plant Mol. Biol. (1989) 13:53-68
                 #title Rice alcohol dehydrogenase genes: anaerobic induction,
                        organ specific expression and characterization of cDNA
                        clones.
                 #cross-references MUID:93357433
                 #accession JQ0474
                 ##status translation not shown
                 ##molecule_type mRNA
                 ##residues 1-376 ##label XIE
                 ##cross-references GB:X16296; NID:g20164; PIDN:CAA34363.1;
                        PID:g20165
                 ##experimental_source seedling, strain IR26
GENETICS
                 #gene Adh1
COMPLEX          dimer
FUNCTION
                 #description catalyzes the oxidation of primary and secondary alcohols
                        to aldehydes and ketones, respectively, by NAD+
                 #pathway alcohol degradation
CLASSIFICATION   #superfamily alcohol dehydrogenase; long-chain alcohol
                        dehydrogenase homology
KEYWORDS         alcohol metabolism; dimer; metalloprotein; NAD;
                 oxidoreductase; zinc
```

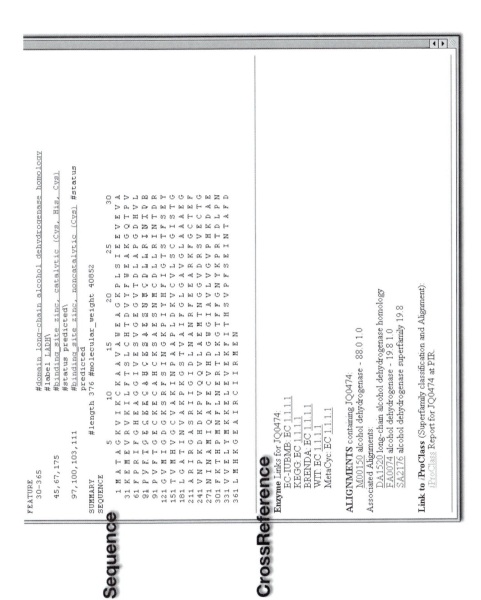

Figure 3.1. An example of a PIR-Protein Sequence Database entry. Note that there are three distinct sections of the entry: annotation, sequence, and cross-references.

Research Foundation (NBRF). The NBRF has been responsible for the primary resource for protein sequences since its inception in 1965 through today's current version, the PSD. The PIR-International was established in 1988 to distribute the PSD internationally. The collaboration brings together experts from the NBRF, the Munich Information Center for Protein S equences (MIPS), and the Japan International Protein Information Database (JIPID). This collaborative effort not only distributes the PSD but it continuously improves it.

PSD Data

A PSD data file is presented in Figure 3.1. The file consists of three parts: annotation, sequence, and classification. The annotation section is the largest as it contains information that characterizes the sequence. These characteristics include core data (entry code, date of entry, author) and information on a protein's identity and classification. PSD supplies many cross-references to external sources of information such as NCBI's organism data and NCBI's PubMed literature reference. Also, a pull-down menu at the top of the page allows for similarity searches and other applications to expand upon the existing query.

Sequence data enters PSD primarily from direct submissions by researchers from published literature, or from translations (of naturally occurring wild-type sequences) from the primary nucleic acid sequence databases GenBank, EMBL, and DDBJ. The sequence is given a unique accession number upon entry into the database. It is then merged with other sequences for the same protein, classified, and annotated.

Sequence submissions from researchers can be emailed to PIR or sent on disk through the mail. The sequence must have been determined by direct protein sequencing methods; a translation of a DNA sequence is not acceptable. PIR has four requirements for a submission: (1) the name of the submitter, (2) a release date, (3) a unique name for the protein, and (4) the sequence in single letter amino acid code.

The data set itself is non-redundant, meaning if PSD has two entries corresponding to the same protein of the same species they will merge them creating an entry for a unique protein. In merging protein sequences, PSD creates the most complete sequence possible for that protein, but they also provide you with information for the retrieval of the original sequences. Data used to annotate sequences can be derived experimentally or can be predicted by computational methods. This is noted in the entry and is important since experimental data is considered to be more accurate. Their annotation uses a standard nomenclature and controlled vocabulary, so it is easy to relate data across files as well as perform accurate and comprehensive text searches. Classifying proteins into evolutionarily related groups is a major focus of the annotation effort. All proteins are classified into protein superfamilies upon entrance into PSD. Their objective is to provide full superfamily/family, domain, and motif classification. Family classification is based on end-to-end sequence similarity. If 20 to 44% of the amino acids in two sequences match then the two proteins are grouped in the same superfamily. It takes 45% or greater end-to-end sequence similarity to place two sequences in the same family. Overall, sequence annotation is classification-driven and rule-based, performed and reviewed by experts, resulting in consistent and accurate data describing a sequence.

Finding Sequences of Interest in the PSD

Over 200,000 sequence entries can be found in PSD so it is important to know what options are available to find ones of interest and how the options work. You can retrieve a sequence from lists filtered by a sequence's source organism and other various functional, chemical, and physical properties or you can find a sequence by submitting a search. Once you have the sequence, PIR has tools to help you further analyze it. The home page provides direct access to tools for data retrieval, data searches, and data analysis.

The data retrieval items include databases and sequence lists managed at PIR, augmenting PSD, such as the ASDB, NRL3D, RESID, and

ALN databases and the sequence lists *Selection List* and *Complete Genomes*. These resources narrow the search to a defined set of sequences, so you are not dealing with all the publicly available sequences to be found in PSD.

Both *Selection List* and *Complete Genomes* are updated bi-weekly, concurrent with updates made to the PSD. From the PIR home page, click the highlighted link *Selection List* or *Complete Genomes* and a new page opens dedicated to the respective resource. For example, if you wanted to find all publicly available protein sequences for the bacterium *Salmonella ordonez*, you could go to the PIR home page (URL found in Table 3.1) and click on *Selection List*. That will open a new page showing a number of categories specific to different groupings of protein sequences, one of which is "Species". Under "Species", click on "s" for *Salmonella ordonez* and a new page appears with all "s" species with sequence entries in the PSD. Clicking on *Salmonella ordonez* will provide you with the sequence IDs for all sequences available (which can then be clicked to see the actual file in PSD). Finding a sequence of interest in *Selection List* may be easier for you than entering a species name in a keyword search of the entire PSD because, though you must navigate through a few pages, they are well organized and almost entirely text, so it is straight forward and fast. In effect, *Selection List* has already done a series of keyword searches for you and organized the results alphabetically.

Complete Genomes takes the "Species" category one step further by presenting a list of proteins that entails a full-complement of all the proteins in a genome, which is the proteome. Currently only the smaller genomes of microbes, mitochondrian, plastids, and chloroplasts are available.

The PIR provides a variety of ways to directly search for and analyze sequences in the PSD. The search tools allow you to paste in a protein sequence or enter in a PIR-PSD code for a sequence and find similar sequences. The similarity searches can be a complete (global or end-to-end) sequence search or a partial (domain/motif, pattern, peptide) sequence search. BLAST or FASTA searches are provided for complete sequence searches. The partial sequence searches are named corresponding to the type of search. For example, *Peptide Match* scans the PSD for exact matches to a partial protein (peptide) sequence.

PSD Sequence Analysis Tools

If you have compiled a list of sequences, similar to the one you provided, by doing a sequence similarity search, you can then align all of the sequences or sort them according to annotation data using PIR's analysis tools. An example here would be if you entered your own sequence and were returned numerous similar sequences, you might want to align the results to look for conserved regions or sort them by taxonomy.

IESA, Integrated Environment for Sequence Analysis, is a tool that allows you to display data (i.e., protein family, domains, species, keywords, etc.) for proteins in a visually appealing and concise table. Opening the IESA tool, you enter a table that consists of all protein sequence entries in PSD, which is over 4,000 pages long. You can pare down the table to entries of interest through searches provided at the top of the page. For example, from the search pull-down menu, select *species* and type in *Salmonella ordonez*, press *go*, and the table then displays sequence entries for that species only. Because annotation data for all of those proteins is presented on a single page, you can compare them without flipping back and forth between a number of pages. Now you have various analysis options for any or all of the sequences in the table. For example, if you want to align them using the CLUSTALW program, select all the sequences then select *multiple alignment* and press *go*.

IESA is very similar to the PIR home page because you can access the same resources from both of them. IESA just has a different interface to the same data, which might be easier and add more functionality to analyses.

Lastly, GeneFind and ProClass identify protein sequence families through the integration of classification data.

SWISS-PROT

SWISS-PROT is the other major protein sequence database, comparable to PIR-PSD in many ways. Though they do not collaborate, users have access to the same set of proteins with similar ways to get them and similar tools to analyze them.

SWISS-PROT is a database with a collection of protein sequences. SWISS-PROT groups at the Swiss Institute of Bioinformatics (SIB) and the European Bioinformatics Institute (EBI) have collaboratively maintained the database since 1987, a year after its inception. It is accessible through the Expert Protein Analysis System (ExPASy) server made available by the SIB (see Table 3.1 for URLs). ExPASy itself is an outstanding site with many services for those interested in all aspects of proteins (see Chapter 11 for a description of their *Proteomics Tools* section) and the applications of SWISS-PROT are extensive due to this allegiance.

SWISS-PROT Data (and the reason for TrEMBL)

The data file is similar to a PSD file in content and structure. Annotation data, sequence data, and cross-references to other online resources provide a thorough explanation of the protein of interest. You will find the core entry data, classification information (sequence families and domains), structural data, and functional data in the annotation. The *Feature table* viewer for each entry provides a graphical display of the sequence with various characteristics associated with annotation data mapped onto it.

SWISS-PROT supplies in-depth documentation on how protein sequence entries are annotated. Knowing how they fill out an entry can be very helpful in assessing data and finding similar data. It also would be of interest anyone interested in a career in bioinformatics. You can find this documentation on the SWISS-PROT home page under *SWISS-PROT documents*. In summary, annotation data primarily comes from published literature, from sequence entries in SWISS-

PROT of evolutionarily related proteins, and from computer programs/ algorithms. Protein family data is kept up-to-date through interaction from SWISS-PROT with experts whose focus is a specific protein family. By no means is annotation a "copy and paste" process. The information is thoroughly assessed before being included in the entry.

Data Sources

SWISS-PROT sequence data comes from submissions by researchers who have experimentally sequenced proteins, and by nucleic acid translations incorporated from TrEMBL. TrEMBL, a database of translated nucleic acid sequences, acts as a supplement to SWISS-PROT. The individual attention to the annotation of protein sequences in SWISS-PROT causes a bottleneck in the rate new proteins enter it. TrEMBL was created in 1996 as a means to keep current and comprehensive without compromising the quality of the manually annotated SWISS-PROT entries. This means users have access to all publicly available protein sequences in any search, but some will have more thorough annotation (SWISS-PROT) than others (TrEMBL). The databases effectively act as a single entity since searches act on both sets of data, not one or the other. The entry that is returned from a search clearly indicates whether the user is looking at a SWISS-PROT or TrEMBL entry.

TrEMBL, Translation of the EMBL nucleotide sequence database, consists of all translated nucleic acid protein coding sequences available in EMBL, except those proteins already integrated into SWISS-PROT. TrEMBL proteins are automatically annotated. There are over 400,000 TrEMBL entries to nearly 100,000 SWISS-PROT entries. A subset of TrEMBL entries will be incorporated into SWISS-PROT (SP-TrEMBL). REM-TrEMBL consists of entries that have no accession numbers and are not set to go into SWISS-PROT. For a sequence to enter SWISS-PROT, it must receive full manual annotation. Again, for more details about this process, see the documentation available through SWISS-PROT.

Table 3.1. Primary Protein Sequence Databases

#	Site Name	URL
1.	PIR-Protein Sequence Database	http://pir.georgetown.edu/
2.	SWISS-PROT + TrEMBL	http://ca.expasy.org/sprot/
3.	OWL	www.bioinf.man.ac.uk/dbbrowser/OWL/
4.	Entrez Protein (NCBI)	http://www.ncbi.nlm.nih.gov/entrez/query.fcgi
5.	PRF/SEQDB	http://www.prf.or.jp/en/

Related Sites

#	Site Name	URL
6.	MIPS - Munich Information Center for Protein Sequences	http://mips.biochem.mpg.de/
7.	JIPID -Japan International Protein Information Database	http://pir.georgetown.edu/pirwww/aboutpir/collaborate.html
8.	IESA - Integrated Environment for Sequence Analysis	http://pir.georgetown.edu/pirwww/search/piriesa.shtml
9.	Swiss Institute of Bioinformatics (SIB)	http://www.isb-sib.ch/
10.	European Bioinformatics Institute (EBI)	http://www.ebi.ac.uk/
11.	Expert Protein Analysis System (ExPASy	http://www.expasy.ch/
12.	NBRF - National Biomedical Research Foundation	http://pir.georgetown.edu/nbrf/
13.	Human Proteomics Initiative (HPI)	http://www.expasy.ch/sprot/hpi/
14.	HAMAP - High quality Automated Microbial Annotation of Proteomes	http://www.expasy.ch/sprot/hamap/

SWISS-PROT/TrEMBL Search and Download

SWISS-PROT/TrEMBL is searched through common fields: accession, ID, citation (SWISS-PROT only), author, full-text, and SRS (Sequence Retrieval System). *Swiss-Shop* is an email update resource for those interested in receiving the latest sequence types to enter SWISS-PROT from TrEMBL or SWISS-PROT/TrEMBL from submissions from researchers. Email updates are sent weekly. SWISS-PROT and a non-redundant combination of SWISS-PROT/TrEMBL can be downloaded via FTP or delivered on CD-ROM. All of this is accessed through the SWISS-PROT home page.

Other Features: HPI, HAMAP

Those who maintain SWISS-PROT started HPI, the Human Proteomics Initiative, in 1999. The project goal is to annotate all protein sequences of the human genome. This entails first determining all coding sequence regions on the genome, then annotating the proteins for function, post-translational modifications, domain make-up, subcellular location, variants, etc. These are the same requirements for proteins found in SWISS-PROT. The SIB and EBI are looking for support from the scientific community in this process.

HAMAP is the High quality Automated Microbial Annotation of Proteomes. As of the summer of 2001, 38 microbial proteomes are in the process of becoming fully annotated. The *Current Status of HAMAP* shows the progress of proteome development. This is a very good resource for those interested in proteomes and genomes of microbes.

Other Protein Sequence Databases

OWL

OWL is a collection of non-redundant protein sequences. OWL data sources are SWISS-PROT (+ TrEMBL), PIR, GenBank (translation), and NRL-3D. SWISS-PROT is the defining source and all others are compared against its contents to eliminate identical sequences. OWL is developed by the same team that develops the protein classification database Prints and the sequence alignment viewer Cinema.

Non-redundant data sets are good sets against which to perform similarity searches because the results, a list of sequences, will not be burdened with excess. OWL has nearly 280,000 entries, similar to the number of non-redundant entries found in PIR-PSD. Different groups attempting the same thing will generate different results.

OWL offers basic search tools, and the result of a search is a list of sequences with links of individual entries sending the user to SWISS-PROT/TrEMBL data files. In effect, OWL is a tool that created a non-redundant, comprehensive data set and search tools to access it. It does not store any data.

Entrez Protein Database

Entrez is a database query tool (see Chapter 10 for more information) with a section designated to proteins. Entrez' protein database compiles protein sequence data from SWISS-PROT, PIR, PRF, PDB, and GenBank nucleotide sequence translations among others.

Peptide/Protein Sequence Database (PRF/SEQDB)

The Protein Resource Foudation's Peptide Institute in Japan created the Protein Sequence Database: PRF/SEQDB. The database contains sequence entries of peptides and proteins as well as translations of

DNA sequence. It has some sequences not found in the other major protein sequence databases because it contains all sequences found in the literature. It currently has nearly 140,000 entries.

Further Reading

Bairoch, A., and Apweiler, R. 2000. The SWISS-PROT protein sequence database and its supplement TrEMBL. Nucleic Acids Res. 28: 45-48.

Barker, W.C., Garavelli, J.S., Hou, Z., Huang, H., Ledley, R.S., McGarvey, P.B., Mewes, H., Orcutt, B.B., Pfeiffer, F., Tsugita, A., Vinayaka, C. R., Xiao, C., Yeh, L.L., and Wu, C. 2001. Protein Information Resource: a community resource for expert annotation of protein data. Nucleic Acids Research. 29: 29-32.

Junker, V.L., Apweiler, R., and Bairoch, A. 1999. Representation of functional information in the SWISS-PROT data bank. Bioinformatics. 15:1066-1067.

Chapter 4

Secondary Nucleotide Databases

Contents

Abstract

Secondary nucleotide databases pull specific types of data from the primary nucleotide databases, GenBank, EMBL, and DDBJ, in generating a subject specific set of data. They offer extensive resources to the subject they cover including background information, pertinent literature, and more thoroughly annotated sequences. These databases cover all realms of nucleotide sequence: uRNA, tmRNA, ESTs (Expressed Sequence Tags), STSs (Sequence Tagged Sites), plasmids,

From: *Genomes and Databases on the Internet: A Practical Guide to Functions and Applications*
ISBN 1-898486-31-X © 2002 Horizon Scientific Press, Wymondham, UK.

vectors, subviral RNAs, etc. The sites themselves are not very difficult to navigate because of their small size.

Introduction

A secondary nucleotide database is a filtered component of a primary sequence database. Whereas EMBL, GenBank, and DDBJ contain all publicly available nucleotide sequences, a secondary sequence database extracts sequences specific to a particular data set—i.e., small RNA or gene specific DNA—and supports them with a searchable site all their own. These sites are not under one watchful eye so the creativity of the curator decides the site's qualities. This leads to a variety of databases, both in content and structure.

Secondary nucleotide databases are more than a source of sequence data. Their content is focused on a discrete subject lending to a greater depth of information per molecule than in the primary nucleotide sequence databases, GenBank, EMBL, and DDBJ. The information they provide that makes them different than the primary databases includes background materials, references to review articles, extensive links to further information, and a more thorough annotation of the sequence.

The intent of this chapter is primarily to inform the science community of the existence of the smaller nucleotide databases on the web. The databases are easily found through a search, which is explained in greater detail below. Tables 4.1 and 4.2, listing DNA and RNA databases respectively, show a few examples of the type of databases to be found. Below are general comments on what to expect when entering one of these sites, both in terms of what they contain and how to use them.

Table 4.1. Secondary DNA Databases

#	Database	URL
1.	Eukaryotic Promoter Database	www.epd.isb-sib.ch/
2.	Codon Usage Database	www.kazusa.or.jp/codon/
3.	EGAD (human DNA)	www.tigr.org/tdb/egad/egad.html
4.	GENOTK (human cDNAs)	http://genotk.genome.ad.jp/
5.	UniGene Resources (gene oriented clusters)	www.ncbi.nlm.nih.gov/UniGene/index.html
6.	TBASE (knockouts)	www.bioscience.org/knockout/dualindx.htm
7.	TransTerm (start/stop codons)	http://uther.otago.ac.nz/Transterm.html
8.	dbEST (expressed sequence tags)	www.ncbi.nlm.nih.gov/dbEST/index.html
9.	dbSTS (sequence tagged sites)	www.ncbi.nlm.nih.gov/dbSTS/index.html
10.	VectorDB	www.atcg.com/vectordb/
11.	NCCB VECTOR DATABASE	www.cbs.knaw.nl/nccb/
12.	NCCB BACTERIA/PLASMIDS DATABASE	www.cbs.knaw.nl/nccb/
13.	NCCB GENE LIBRARIES	www.cbs.knaw.nl/nccb/

Finding DNA and RNA Secondary Databases

Before discussing the finer points of secondary sequence databases, you should know that if you have a specific data set in mind it would be wise to do a web search to see if a database for it exists. The variety of small sequence databases is great and ever expanding, meaning many seemingly obscure subjects have their own public database.

Table 4.2. Secondary RNA Databases

#	Database	URL
1.	5S Ribosomal RNA	http://biobases.ibch.poznan.pl/5SData/
2.	Aminoacyl-tRNA Synthetases Database	http://biobases.ibch.poznan.pl/aars/
3.	Database of Non-coding RNAs	http://biobases.ibch.poznan.pl/ncRNA/
4.	Aptamer database	http://rocko.icmb.utexas.edu/APTAMER/
5.	Small RNA database	http://mbcr.bcm.tmc.edu/smallRNA/smallrna.html
6.	The RNA Modification Database	http://medlib.med.utah.edu/RNAmods/
7.	The distribution of RNA motifs in natural sequences	www.centrcn.umontreal.ca/~bourdeav/Ribonomics/
8.	snoRNA Database	http://rna.wustl.edu/snoRNAdb/
9.	The uRNA Database	http://psyche.uthct.edu/dbs/uRNADB/uRNADB.html
10.	UTRdb: a database of Untranslated Regions of Eukariotic mRNAS	http://bigarea.area.ba.cnr.it:8000/BioWWW/#UTRdb
11.	The tmRNA Website	www.indiana.edu/~tmrna/
12.	PseudoBase	wwwbio.leidenuniv.nl/~batenburg/PKB.html
13.	Plant Mitochondrial tRNA Database	http://bio-www.ba.cnr.it:8000/BioWWW/#PLMItRNA
14.	ssu rRNA database	http://rrna.uia.ac.be/ssu/
15.	lsu rRNA database	http://rrna.uia.ac.be/lsu/
16.	Subviral RNA Database	http://nt.ars-grin.gov/subviral/

The three ways to find a database are through a search, a web portal, or a *Related Links* page on similar site. There are so many search engines and varying qualities among them that it is easy to believe that results from each would be different. This is true when looking for a field of general public interest, but with the field of science the search is highly specific. This makes the search more likely to generate accurate results. Google, www.google.com, is a simple, yet thorough search engine. Typing in the name of what you are looking for plus a keyword such as *database* significantly improves the search. For someone more familiar with scientific offerings on the web, you probably know of a science portal with lists of databases. Chapter 12 lists a few portals for various subjects including RNA. Lastly, all molecular biology sites have a page with links to similar sources. The focus of the lists is always on the type of site where it is found: other RNA resources are found at a RNA database.

Site Content

Secondary nucleotide sequence databases generally contain a thorough introduction into the topic, sequence entries, search tools, references, and links to similar sites.

The number of sequences relevant to the subject of interest should be the same as would be found at a primary database (GenBank, EMBL, DDBJ). Several of the DNA databases in Table 4.1 use GenBank as their primary source of data. They scan GenBank for sequences specific to a particular subject and add them to their database. Often the secondary databases add data to that found in the primary data file. The core data fields (accession number, reference, name, etc.) and the sequence will be present at the very least. Some times a data file will be presented in the identical format as seen in the primary database. Generally more functional information is presented.

Site Use

Secondary database sites are simple to use. Their specificity and small size leaves less room to confuse you. There will not be an overwhelming number of options on the home page. At this point you select a specific organism or sequence type or simply scroll through a list to find the sequence of interest. The results come quickly.

The search option is generally quite simple. The more complex searches include parameters with which to apply limitations on a search. Each parameter is explained within the site and is many times accompanied by references for further insight. This information can be very useful to someone new to a field and new to database searches in general.

Secondary Database Examples

Tables 4.1 (DNA) and 4.2 (RNA) contain database names and URLs for a few secondary nucleotide sequence databases on the web. There are more than the list provides, but that is to be expected since these databases are continually being created. The names of the databases give a good indication as to what can be found in this field. Hopefully, it provides interested researchers with the impetus to search for some that are not listed, but may be available.

Further Reading

Bourdeau, V. Ferbeyre, G.,Pageau, M.,Paquin, P., and Cedergren, R. 2000. The distribution of RNA motifs in natural sequences. Nucleic Acids Res. 27: 4457-4467.

Brown, C.M., Dalphin, M.E., Stockwell, P.,and Tate, W.P. 1993. The translational termination signal database. Nuc. Acids Res. 21: 3119-3123.

Graziano, P., Sabino, L., Giorgio, G., Matilde, I., Alessandra, L., Wojtek, M., and Cecilia Saccone, C. 1999. UTRDB: a specialized database of 5' and 3' untranslated regions of eukaryotic mRNAs. Nucleic Acids Research. 27.

Lowe, T.M., and Eddy, SR.A. 1999. computational screen for methylation guide snoRNAs in yeast. Science. 283: 1168-1171.

Luigi, C.R., Volpicella, M., Liuni, S., Volpetti, V. Licciulli, F., and Gallerani, G. 1999. PLMItRNA, a database for higher plant mitochondrial tRNAs and tRNA genes. Nucleic Acids Research. 27: 156-157.

Nakamura, Y., Gojobori, T., and Ikemura, T. 2000. Codon usage tabulated from the international DNA sequence databases: status for the year 2000. Nucl. Acids Res. 28: 292.

Périer, R.C., Praz, V., Junier, T., Bonnard, C., and Bucher, P. 2000. The Eukaryotic Promoter Database (EPD). Nucleic Acids Res.28: 302-303.

Singer, M. *et al.* 1989. A collection of strains containing genetically linked alternating antibiotic resistance elements for genetic mapping of *Escherichia coli*. Microbiol. Rev. 53: 1-24.

Szymanski, M., Miroslawa, M., Barciszewski, Z., Volker, and J. Erdmann, A. 2000. 5S ribosomal RNA database Y2K. Nucleic Acids Res. 28: 166-167.

Szymanski, M., and Barciszewski, J. 2000. Aminoacyl-tRNA synthetases database Y2K. Nucleic Acids Res. 28: 326-328.

Volker, A. Erdmann, M., Szymanski, A.,Hochberg, N., and Barciszewski, J. 2000. Non-coding, mRNA-like RNAs database Y2K. Nucleic Acids Res. 28:197-200.

Zwieb, C. 1996. The uRNA database. Nucleic Acids Research. 24: 76-79.

Chapter 5

Protein Classification Databases

Contents

Abstract
Introduction
Common Attributes
Individual Descriptions

From: *Genomes and Databases on the Internet: A Practical Guide to Functions and Applications*
ISBN 1-898486-31-X © 2002 Horizon Scientific Press, Wymondham, UK.

Abstract

Classifying proteins organizes them with respect to similarity. Proteins are placed into groups of similar proteins, usually called a protein family. There are online databases dedicated to protein classification. There are several of these databases because there are different methods of classification. The methods vary depending on how proteins are compared: sequence versus structure, global alignment versus local alignment, and manual efforts versus automated efforts. Some databases integrate multiple databases into their results and others classify proteins based on their own method. Whatever the case, they are all slightly different, but all quite informative. Determining which family a protein belongs to can go a long ways towards defining its function.

Introduction

Protein classification databases group, or classify, proteins based on sequence similarity, making them useful in defining the function of an unknown protein. A similar sequence generally means a similar structure, which often means a similar function. If a researcher has a sequence of an unknown protein, these databases would be a good place to start an initial analysis of its function because there may be a similar protein in the public domain whose function has been determined. Like every other data source on the web, protein classification databases take advantage of experimentation that has already been done. These databases can give you a good idea as to an unknown gene's role in an organism.

The majority of protein classification databases use sequence similarity in arranging proteins into groups. The few databases that classify proteins by structural similarity are described in Chapter 6, Molecular Structure Databases. URLs to all protein classification sites, including those of structure-based classification, are found in Table 5.1.

The following includes a general description of the field of protein classification databases (the process of classification and their utilization of other databases) and reviews each database one by one.

Common Attributes

Words such as motif, pattern, profile, cluster, domain, signature, family, and groups are used to describe the way proteins are related by similarity. They mean slightly different things but they refer to the same goal: classification. The descriptions below use family for a group of related proteins and domain for a conserved sequence segment.

Classification Schemes

Proteins can be classified in different ways. Aligning sequences against each other is the foundation of classification. From there it gets complicated.

Most classification schemes take into account local alignments. The obvious and easiest way to classify sequences is to compare them from start to finish (global alignment) and arrange them into families by overall percent similarity. This method is not very thorough when considering there are regions of significance interspersed within a complete sequence. A region of significance refers to a sequence segment that is often specific to a function or other property of the

Table 5.1. Protein Classification Databases

#	Database Name	URL

By Protein Sequence

#	Database Name	URL
1.	iPROCLASS	http://pir.georgetown.edu/iproclass/
2.	InterPro	http://www.ebi.ac.uk/interpro/
3.	MetaFam	http://metafam.ahc.umn.edu/
4.	COGs	http://www.ncbi.nlm.nih.gov/COG/
5.	PROSITE	http://www.expasy.org/prosite/
6.	Pfam	http://pfam.wustl.edu/
7.	Blocks	http://www.blocks.fhcrc.org/
8.	ProtoMap	http://www.protomap.cs.huji.ac.il/
9.	PRINTS (PRINTS-S)	http://bioinf.man.ac.uk/dbbrowser/PRINTS/
10.	ClusTr	http://www.ebi.ac.uk/clustr/
11.	eMOTIF	http://motif.stanford.edu/emotif/
12.	SBASE	http://www3.icgeb.trieste.it/~sbasesrv/
13.	TIGRFAMS	http://www.tigr.org/TIGRFAMs/
14.	DOMO	via http://www.infobiogen.fr
15.	ProDom	http://protein.toulouse.inra.fr/prodom/

By Protein Structure*

#	Database Name	URL
16.	HOMSTRAD (+ PLUS)	http://www-cryst.bioc.cam.ac.uk/data/align/
17.	FSSP	http://www2.ebi.ac.uk/dali/

18.	CAMPASS	http://www-cryst.bioc.cam.ac.uk/~campass/
19.	SCOP	http://scop.mrc-lmb.cam.ac.uk/scop/
20	CATH	http://www.biochem.ucl.ac.uk/bsm/cath/
21.	PDB-REPRDB	http://www.rwcp.or.jp/papia/

* These sites are reviewed in Chapter 6.

protein, for instance the binding of protein to DNA. If this sequence segment—the domain—is found in two sequences, they are functionally similar. Their binding to DNA narrows the possibilities of what their function might be. Databases may also treat local alignments differently, either by basing their families on a single domain or by incorporating the domain make-up (profile or fingerprint) of the entire sequence in generating families. Families in Blocks, PRINTS and PROSITE, for example, take into account multiple domains, and families in Pfam, ProDom and DOMO correspond to individual domains. Documentation at the websites of these databases explains their intricacies in more detail.

Manual vs. Automated

Classification schemes are either manual or automated. They are all automated to some degree, but some are entirely automated, whereas others, such as PROSITE, incorporate a little human touch. In general, an expert that checks the results of an automated process and makes adjustments to obvious errors adds quality to the classification. Many times, however, a researcher does not need or is not looking for an answer to a direct question ("What is the function of this gene?"), and instead wants to perform a complete genome analysis. In that case, a few automated "errors" are not catastrophic. Databases such as CluSTr are geared for genome/proteome scale analyses.

Database Integration

Integration of various data sources is a way to add quality to a data resource. This comes in two forms. At the results page for a protein, where one would find a list of all the proteins in their group and a description of function common to all of them, a link to another database's results for the same protein will be given. This makes it possible to quickly compare a result. The other way to integrate databases is for the curators to incorporate other database results in their classification scheme. For example, a curator might make a consensus of protein families from all other databases. In effect, the database curator is doing the comparison of different database results for the user and merging it into a single report.

In regards to database integration, the primary protein sequence databases, PIR-PSD and SWISS-PROT, as well as the Protein Data Bank (structure), provide links to many of the protein classification databases. Their cross-references are as extensive or more so than all of the classification databases, making them the place to start in any search. There are unique services offered by most of the classification databases, from a phylogenetic tree representation of the members of a protein family to an alignment of multiple sequences for a conserved region. This makes them useful beyond their function as a cross-reference from a primary database.

Downloading Files

If a user's plan is to analyze a batch of sequences, it is best to download the entire database to your personal computer to save time. When you download, you don't have to send data through the internet and wait for its return each time. Most databases offer this for free to academic users through FTP downloads (see Chapter 1 for an FTP download explanation).

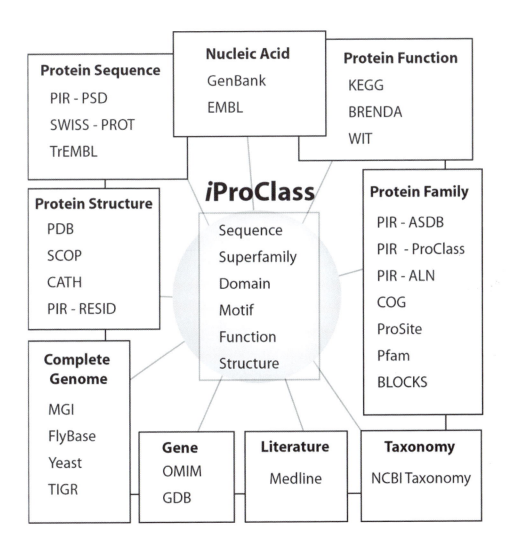

Figure 5.1. The data integrated by iPROCLASS, an integrated protein classification database, includes protein sequences, structures, and families, nucleotide sequences, literature references, taxonomy, etc.

Individual Descriptions

iPROCLASS

iPROCLASS is an extension of PROCLASS, the Protein Information Resource's (PIR) first attempt at integration of classification data. PROCLASS integrated PIR-PSD superfamily data with PROSITE data. IPROCLASS extends beyond classification databases by integrating structure and function databases (see Figure 5.1). It must be said that other classification databases do not strictly show the evolutionary associations between proteins but provide extensive annotation of protein functions (e.g., InterPro and PROSITE to name a few).

The results of the integration are not merged. In effect, iPROCLASS submits a search to all the databases and compiles a list of sorted links to results at those databases.

InterPro

InterPro, Integrated resource of Protein Families, Domains and Sites, is a database that combines the results of other databases into a single report. The databases it integrates are Pfam, PROSITE, PRINTS, ProDom, SMART, and SWISS-PROT/TrEMBL. InterPro does not simply throw different data sets onto a single page with their individual formats and nomenclatures. Instead, it merges data in providing a concise yet comprehensive well-formatted text report. A weakness in one database is complemented by strengths in others. InterPro is a refreshing database because of its streamlined approach (users can only handle so much text on screen). Links to the individual databases allows a user to quickly jump to the original files if so inclined.

The future of InterPro involves incorporating TIGRFAMs, as well as other classification databases.

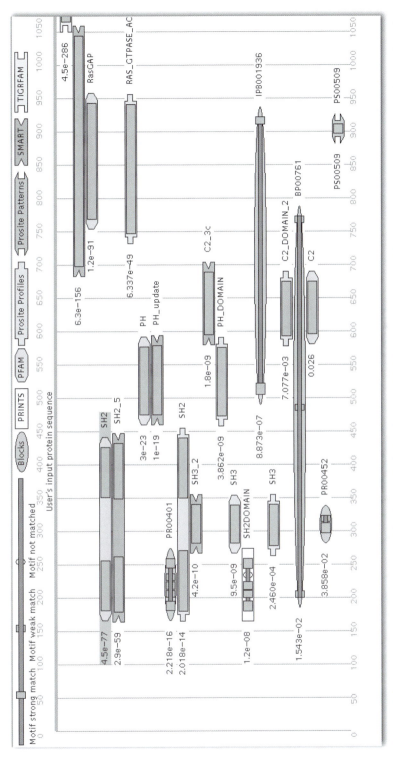

Figure 5.2. An example of results from MetaFam's PANAL (Protein ANALysis tool) system. The results show the differences in classification among the databases MetaFam integrates. Different domain icons are associated with the different classification databases that generated them.

MetaFam

This database is designed to integrate the various classification databases that exist. Results include a graphical display (*MetaFamView*) of the interrelationship between entries among the different databases: PROSITE, Blocks, PRINTS, and Pfam. Another display represents domain locations on a protein sequence as determined by the integrated set of databases (see Figure 5.2). MetaFam supports cross-references in the form of links to the individual reports.

COGs

Clusters of Orthologous Groups (COGs) like TIGRFAMs, makes a strong attempt to function as the driving force in grouping proteins. Organizing proteins in orthologous (or paralogous) groups across species represents this effort. This results in every protein in a COG having a common ancestral protein and every function deriving from a common function.

Only complete genomes are compared in this process (see documentation at website for further information), which makes this database good for studying evolutionary patterns between species. This sets it apart from other classification databases because it goes beyond molecule-by-molecule research and effectively classifies organisms.

Because NCBI manages this database, many features NCBI offers are cross-referenced in the results, making vast amounts of information available for each protein.

PROSITE

PROSITE is a database of protein families and domains. Its focuses on biologically significant regions in protein sequences. These include

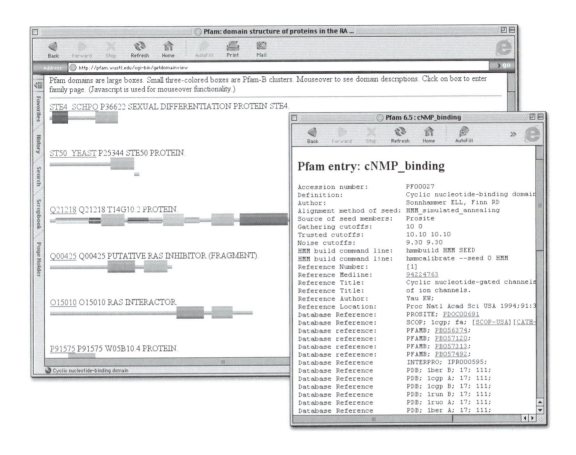

Figure 5.3. An example of results from the Pfam protein classification database. The image in the background is a graphical display of domain location on protein sequences. The image in the foreground represents data acquired by clicking on one of the domains from the background image.

catalytic sites, prosthetic group binding sites, metal ion binding sites, molecular (i.e., ATP, DNA, calcium, other proteins) binding sites, and sites involved in disulfide bonding. It is accessible through the ExPASy server, so it is tightly linked to all resources available there including SWISS-PROT and its expert annotation. Data on proteins and families observed through PROSITE is as exhaustive as any classification database. PROSITE takes into account the multiple domains of each sequence when generating its families.

Searches reveal protein families whose function are as fully described as public knowledge permits. A general description of the family and the individual domains comprising it (as consensus domain sequences) is the backbone of the data.

Tools to compare an unknown sequence to the families and domains in PROSITE are available (ScanProsite and ProfileScan).

Pfam

Pfam looks at all domains in each sequence and generates families as proteins that correspond to a single domain. Curators of the database do this through manual alignment, generating high quality results. Because the manual approach takes time and Pfam would like to represent as many proteins they can without sacrificing quality, they have labeled those proteins that have undergone manual alignments as Pfam-A data. Pfam-B data consists of all other data, which has been classified by an automated approach. Another difference between A and B data is that A has been annotated with protein function data whereas B is not annotated. The entire data set represents all sequences in SWISS-PROT.

Pfam supports classification analysis of unknown sequences (sequence not represented in SWISS-PROT), using basic search tools. Results are presented graphically (see Figure 5.3) with links to additional text-based data. Colored blocks (thick for Pfam-A, thin and multi-colored for Pfam-B) represent the domains. Running the mouse over the domain gives its name at the bottom of the browser. Clicking on a domain leads to more data in the form of cross-references to both sequence and structure classification databases.

Blocks

The Blocks server supports a data set of conserved domains (blocks) from aligned protein sequences. The aligned set of sequences, protein families, comes from InterPro. Blocks families are based on a comparison of protein's domains, not a single domain among many others. This leads to the Blocks database containing over 4,000 conserved domains and nearly 1000 protein families.

A search at the Blocks server does not, however, have to be done solely on the set of data stored in the Blocks database (4,000 domains, 1,000 families). It is supplemented through integration with other protein classification databases—PRINTS, Pfam, PRODOM, and DOMO—effectively making it a search covering 11,000 domains from 2,400 families.

A Blocks search yields one or more conserved domains within a protein family. The name of the protein family is given along with a link to that family's InterPro web interface that provides the user with information on its function. Each protein is directly linked to SWISS-PROT, which provides information on the protein's function and other information. From the SWISS-PROT entry, the user can find links to various other databases for more information.

The Blocks server also allows a researcher to use its algorithms (Block Maker) to find conserved domains in their own protein (and DNA) multiple sequence sets.

ProtoMap

ProtoMap classifies all SWISS-PROT protein sequences through an automated approach. Every protein from SWISS-PROT is classified into "clusters," despite the level of relatedness. This does not mean ProtoMap is careless; the hierarchical approach to classification includes values of relatedness to illustrate how closely related individual proteins are to each other.

Search results include information for both individual sequences and clusters. Links to the SWISS-PROT entry for a sequence and the PROTOMAP cluster group are provided for each protein revealed in a search. The cluster groups are presented as individual sequences. A link to the PROSITE description of the cluster is provided.

This Java oriented database produces two very interesting graphical displays: one is a phylogenetic tree of every sequence in a single cluster and the other displays cluster-to-cluster relatedness. The displays are interactive, so you can link to data for any of the proteins or clusters as you pass over them with the mouse.

PRINTS

PRINTS refers to fingerprints—an identity of a protein based on its entire assortment of domains. Prints creates protein families by looking at all the domains in a sequence, not on whether the sequence has a single domain. The PRINTS data source is SWISS-PROT/TrEMBL.

PRINTS results include a thorough description of the family, literature references, cross-references (PROSITE, Blocks, Pfam, InterPro), individual protein links to SWISS-PROT, and a very easy to understand view of the motif (domain) elements making up the family. *View Relations* (biologically similar families) and *View Alignments* (Cinema alignment viewer) are two other interesting features PRINTS provides.

CluSTr

CluSTr, developed by the SWISS-PROT/PROSITE group, is a database of protein families generated entirely by automated methods. As we discussed in Common Attributes, databases of this nature are most appropriately used for whole genome (or more accurately, proteome) analyses.

eMOTIF

eMOTIF examines multiple sequence motifs, or domains, within each sequence alignment. It focuses on highly conserved domains to avoid inconsistencies. eMOTIF derives its data from sequence alignments in Blocks and Prints. In fact, the results take the user to the Blocks or Prints entry for the protein. See the Blocks and Prints descriptions above for what information that entails.

In addition to a basic search of the public domain, users are able to perform their own domain scan of an aligned set of unknown sequences with the same algorithm eMOTIF used to generate its data set: eMOTIF MAKER.

SBASE

SBASE gathers protein sequences from PIR-PSD, SWISS-PROT, and published literature. Classification of the sequences results in SBASE-A and SBASE-B data. Like Pfam, these different data types refer to the quality of the analysis. Type A has been validated and type B has not. The classification scheme involves more than sequence analysis. It includes information such as structure, function, composition, and binding specificity. The domains are validated by comparison to other classification database results (PROT-FAM, Pfam, InterPro).

The results of the data include cross-references to PROSITE, PRINTS, ProDom, Blocks, Pfame, PDB and OMIM. Results are retrieved through a BLAST search.

TIGRFAMs

TIGRFAMs has a logical take on classification: protein families are organized by function so that the function of every protein in a family is uniform. Sequence alignments are still used, but ultimately the decision is based on function. Determining the function of a protein is the primary goal of genome annotation.

Equivalog is the name given to the data TIGR-FAMS generates. An equivalog is a family of homologous proteins with conserved function. There are over 500 equivalogs in the database.

The annotation of function is extensive and is complemented by a link to PROSITE, another well-annotated classification database.

DOMO

Briefly, DOMO is a database of homologous protein domain families. Several automated sequence analysis steps have been used to create nearly 9,000 protein families from sequences obtained from PIR-PSD and SWISS-PROT. The protein families are generated from the approach that family members only have to have one domain in common.

ProDom

Again, this is another classification database that automatically generates homologous families in this database. Proteins within the family need only one domain in common. Built from sequences obtained through SWISS-PROT and TrEMBL, ProDom cross-references these in its results along with PROSITE, Pfam, InterPro, and the PDB.

Others

Other classification databases exist (i.e., ProtFam), but new resources are constantly arising. Reviewing the above databases should give researchers interested in utilizing classification databases a comprehensive view of what may be encountered on the web.

Further Reading

Apweiler, R., Attwood, T. K., Bairoch, A., Bateman, A., Birney, E., Biswas, M., Bucher, P., Cerutti, L., Corpet, F., Croning, M.D.R, Durbin, R., Falquet, L., Fleischmann, W., Gouzy, J., Hermjakob,

H., Hulo, N., Jonassen, I., Kahn, D., Kanapin, A., Karavidopoulou, Y., Lopez, R., Marx, B., Mulder, N.J., Oinn, T.M., Pagni, M., Servant, F., Sigrist, C. J. A., Zdobnov, E. M. 2001. The InterPro database, an integrated documentation resource for protein families, domains and functional sites. Nucleic Acids Res. 29: 37-40.

Attwood, T. K., Croning, M. D. R., Flower, D. R., Lewis, A. P., Mabey, J. E., Scordis, P., Selley, J. N., and Wright, W. 2000. PRINTS-S: the database formerly known as PRINTS. Nucleic Acids Res. 28: 225-227.

Bateman, A., Birney, E., Durbin, R., Eddy, S.R., Howe, K.L., and Sonnhammer. E.L.L. 2000. The Pfam Protein Families Database Nucleic Acids Res. 28: 263-266.

Corpet, F., Servant, F., Gouzy, J., and Kahn, D. 2000. ProDom and ProDom-CG: tools for protein domain analysis and whole genome comparisons. Nucleic Acids Res. 28: 267-269.

Gracy, J., and Argos, P. 1998. DOMO: a new database of aligned protein domains. Trends Biochem Sci. 12: 495-497.

Haft, D.H., Loftus, B.J., Richardson, D.L., Yang, F., Eisen, J.A., Paulsen, I.T., and White, O. 2001. TIGRFAMs: a protein family resource for the functional identification of proteins. Nucleic Acids Res. 29: 41-43.

Henikoff, J.G., *et al.* 2000. Increased coverage of protein families with the Blocks Database servers. Nucleic Acids Res. 28: 228-230.

Hofmann, K., Bucher, P., Falquet, L., and Bairoch, A. 1999. The PROSITE database, its status in 1999. Nucleic Acids Res. 27: 215-219.

Huang, J.Y., and Brutlag, D.L. 2001. The EMOTIF database. Nucleic Acids Res. 29: 202-204.

Kriventseva, E.V., Fleischmann, W., Zdobnov, E.M., and Apweiler, R. 2001. CluSTr: a database of clusters of SWISS-PROT+TrEMBL proteins. Nucleic Acids Res. 29: 33-36.

Murvai, J., Vlahovicek, K., Barta, E., and Pongor, S. 2001. The SBASE protein domain library, release 8.0: a collection of annotated protein sequence segments. Nucleic Acids Res. 29: 58-60.

Silverstein, K.A.T., Shoop, E., Johnson, J.E., Kilian, A., Freeman, J.L., Kunau, T.M., Awad, I.A., Mayer, M., and Retzel, E.F. 2001. The MetaFam Server: a comprehensive protein family resource. Nucleic Acids Res. 29: 49-51.

Tatusov, R.L., Natale, D.A., Garkavtsev, I.V., Tatusova, T.A., Shankavaram, U.T., Rao, B.S., Kiryutin, B., Galperin, M.Y., Fedorova, N.D., and Koonin, E.V. 2001. The COG database: new developments in phylogenetic classification of proteins from complete genomes. Nucleic Acids Res. 29: 22-28.

Wu,C.H., Xiao, C., Hou, Z., Huang, H., and Barker, W.C. 2001. ProClass: an integrated, comprehensive and annotated protein classification database. Nucleic Acids Res. 29: 52-54.

Yona, G., Linial, N., and Linial, M. 2000. ProtoMap: automatic classification of protein sequences and hierarchy of protein families. Nucleic Acids Res. 28: 49-55.

Chapter 6

Molecular Structure Databases

Contents

From: *Genomes and Databases on the Internet: A Practical Guide to Functions and Applications*
ISBN 1-898486-31-X © 2002 Horizon Scientific Press, Wymondham, UK.

Nucleic Acid Structure Databases
  Nucleic Acid Structure Types
  NDB (& PDB)
  RNABase
IMB Jena Image Library and others
Further Reading

Abstract

Sequence data holds only so much information, especially when it comes to proteins. Protein structures, 3-D structures, are representative of the molecule as it functions in the cell. Knowing what a molecule looks like at its biologically active state is a powerful piece of information. Molecular structure databases are portals into the three dimensional configurations of molecules. Structure databases, primarily concerned with proteins, are used for functional and evolutionary studies of molecules. Databases developed for structural studies of DNA and RNA also exist, though the amount of structural they contain is minimal.

Introduction

A molecule's structure offers significant clues about its function and evolutionary development. This chapter examines the varying databases that feature structural information with an emphasis on protein structures, since nearly all structure data is for proteins.

A 3-D structure is an attempt to look at a molecule as it exists *in vivo*. Because of the influx of sequence data, structural genomics groups have been initiated in an attempt to define the new genes being discovered. The three dimensional (3-D) structure of a molecule (protein, DNA, RNA, protein-DNA complexes, etc.) is a biologically active state that interacts with itself and its environment based on its intrinsic chemical properties, such as hydrophobic and ionic interactions. Structural genomics initiatives are based on the fact that the structure of a protein will help determine its function. Improved methods in experimentally determining structures and the need to functionally analyze gene sequences has led to substantial funding of groups dedicated to generating structure data.

The analysis of structure databases is divided into three sections: Primary (Protein) Structure Databases, Secondary Protein Structure Databases, and Nucleic Acid Structure Databases. The primary structure databases are similar to the primary sequence databases in that they are the point of input of structural data into the web. Structural data for all proteins is initially placed in these databases while a second set of databases, the secondary databases, either filter the primary data set for a specific data type (i.e., structures determined by X-ray crystallography), re-organize the primary data set with respect to evolutionary relatedness (classification), or further annotate individual files of the primary data set (secondary structure determination).

Of note, three dimensional structure modeling and secondary structure prediction tools are reviewed in Chapter 11, Proteomics Tools.

Primary (Protein) Structure Databases

The Protein Data Bank (PDB) is the focus of this section for good reason: it has existed for three decades and until recently, it has been the only database with a full, up-to-date list of publicly available protein structures. The EBI-MSD, Europe's Macromolecular Structure Database, is the new alternative. The two structure databases openly collaborate and share the same data set.

Because the PDB and EBI-MSD share the same data, they are similar. Much of what is said about the PDB will relate to EBI-MSD. EBI-MSD is reviewed briefly after the PDB with a focus on the differences between the two.

Lastly, both of these databases contain nucleic acid structures, as their data sets officially are described as bearing "macromolecular" structure data, but because the amount of nucleic acid structure data is so small the analysis to follow on the PDB and EBI-MSD focuses on proteins. Nucleic acid structure is described separately at the end of the chapter. Table 6.1 includes web addresses for the PDB and EBI-MSD.

Protein Data Bank

The Protein Data Bank is an international database managed by the Research Collaboratory for Structural Bioinformatics (RCSB). Introduced in 1971, the PDB was managed by Brookhaven National Laboratories until 1998 when the RCSB took over. In effect, there are no dramatic changes. The PDB, a worldwide archive of structural data for proteins, is an outstanding resource for scientific community in general.

General Features of the PDB

PDB contains three-dimensional (3-D) structure data on biological macromolecules, mainly proteins. It is a primary database that contains all publicly available protein structure data. When someone has protein structure data to be put on the web, this is the place to put it. Submitted structures are validated by the management staff of the RCSB in coordination with the author of the structure.

All of what the PDB has to offer can be accessed directly from the home page. From it you can search the database, deposit a new protein structure, download a PDB file, preview upcoming changes of the PDB, explore links to other web resources, and find general information on the PDB in the *About the PDB* section.

There are three ways to search the PDB, which are detailed below. The PDB provides a tutorial on how to deposit the data by explaining formatting options and data validation. Downloading data from the PDB can be done from the home page or from an individual data file via FTP. A CD-ROM containing the full PDB is available quarterly for no cost. The *Preview* link on the home page allows the scientific community to involve itself in PDB development by providing the means to test and comment on proposed new features. The *Links* page has links to approximately 40 structure-related databases, nearly 30 structure classification and analysis sites, structure genomics projects, structure verification sites, and other resources of interest.

The PDB is updated frequently and currently has over 15,000 entries. That number should substantially increase in the near term with the inception of structural genomics initiatives.

PDB Data Retrieval and Format

Each entry of the PDB has its own PDB ID. The simplest way to access data is to enter in the PDB ID at the home page. There are two search options, SearchLite and SearchFields, with which to find a PDB entry if you do not have a PDB ID. SearchLite is similar to the general search engines for the web in that you enter one or more terms into a single field and request results. SearchFields is a more refined search by providing a number of fields to narrow the search. Those fields include keywords, sequence similarity (you paste in a sequence), secondary structural features, and other sequence features.

Data is viewed through the Protein Structure Explorer web interface. This is simply the page format of an individual structure file. Figure 6.1 presents an example and will be referred to in the following description of the data.

The initial page of data summarizes all data in the structure file. The complete set of data for PDB entry is enormous and complex, so the *Summary Information* provides an introduction to the file by only presenting core information such as an entry's title, author, entry date, etc. All other data for the entry is accessible from this page through links to *View Structure, Download/Display File, Structural Neighbors, Geometry, Other Sources,* and *Sequence Details.* The Protein Structure Explorer interface keeps these links available from each successive page as you navigate through the entry's data.

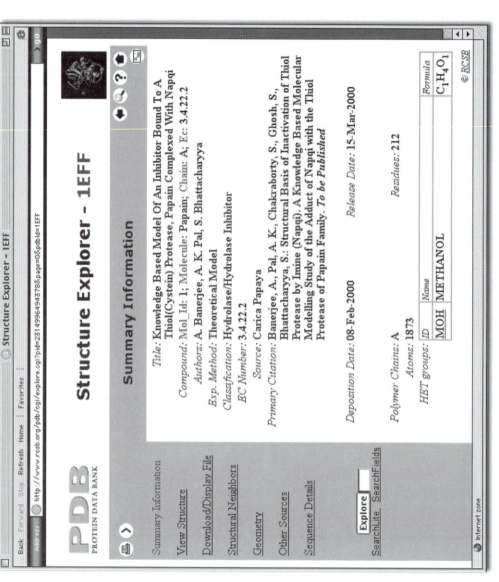

Figure 6.1. An example of a Protein Data Bank entry.

View Structure allows you to view the 3-D protein structure. You can view the structure as a still image or interactively. Still images are displayed immediately with no software downloads necessary. Interactive viewing gives the user the ability to manipulate the image (rotation, magnification, color coding, display style, etc.) Various viewers can be used (see Chapter 11 for more information on viewers). Assistance is provided by the PDB at the *View Structure* page with links to download viewers and help on the subject of viewers.

The *Download/Display File* page presents the raw data of an entry. The files include the 3-D structure data in the form of X, Y, and Z coordinates.

Structural Neighbors is a useful way to access similar structures to a protein. This page links to structural classification databases, described later in the chapter, and structural alignment/comparison tools, found in Chapter 11.

Geometry provides physical characteristics of the protein structure: bond length, bond angle, and dihedral angle data.

Other Sources cross-references the PDB entry to a multitude of other databases and tools to further analyze the protein. When you click on a link to another database or tool a browser window opens up. The PDB code for the protein you are analyzing is already entered into the query table for that resource. You simply initiate the query and the results follow.

The last page of data available for a protein entry in the PDB is *Sequence Details*. Here you will find the sequence for the protein, secondary structure information, and sequences of protein chains.

Data Uniformity and Other Improvements

Data uniformity issues exist because the PDB has been around since 1971 and has undergone improvements in the data entry format. Old PDB entries consisted of one field of text plus the structure. Currently, there are many text fields, or records, per entry in addition to the structure. The different features in an entry allow the PDB to make a more complex search of the data. With the old records you could only query by the PDB ID. The clean up of old data is a collaborative process between the PDB and the EBI-MSD.

The PDB data format has been the PDB file and undergoes continuous changes. The mmCIF file format is being incorporated as an underlying standard that includes CORBA and XML additions. These technical terms were mentioned to show the PDB is always looking to apply the latest in computer programming techniques as a way to improve access to the data. The complex files being generated are more easily transferred across database platforms meaning greater sharing capabilities thus greater access to the scientific community.

PDB wants to improve the process between the deposition of data to its public display. Not only do they want to shorten the period between deposition and display, they also want to increase the speed of transfer of data of a submission to the PDB by creating more deposition sites around the world. Specifically, the EBI-MSD helps in this effort by receiving data from the European community.

In the future, PDB hopes to provide database tools to be used locally, eliminating the need to go through the web. You will be able to download these tools to your personal computer.

EBI-Macromolecular Structure Database

The EBI-MSD site states clearly that its development, initiated in 1996, is not to displace the PDB but to add variety and quality to the resources available to the scientific community interested in protein structure. The fact that the two databases operate on the same data set shows this to be evident.

EBI-MSD provides alternatives to the PDB in terms of data management and services. It acts as a European deposition site of structure data. EBI-MSD asks researchers to deposit data based on proximity to either of the two sites, even though the modes of deposition (*ADIT* for the PDB and *AutoDep* for the EBI-MSD) are different. For example, those in the USA should deposit through the PDB site and those in Europe should deposit through the EBI-MSD site. It takes less time to deposit to a database that is geographically closer to you.

EBI-MSD offers three services not found on the PDB: *Probable Quaternary Structures* (*PQS*), *3Dseq*, and *Unpublished References*. *PQS* supports quaternary protein structures, aggregates of units of tertiary protein structures. These aggregates are biologically active hence they can lead to functional clues about a protein. *3Dseq* removes inconsistencies in the PDB sequences of structures and cross-references them to the SWISS-PROT protein sequence database. This service aids in extracting complete and accurate sequence information for a PDB structure. The *Unpublished References* service lists journal references to structures in the PDB where the publication name is "TO BE PUBLISHED".

EBI-MSD also has a collaborative relationship with BioImage, a site for microscopic images ranging from macromolecules to cells to entire organisms. Features of the EBI-MSD database are not fully understood because it is still in the design stage; however, it will offer data from BioImage, an example of the difference in data management between the PDB and the EBI-MSD.

Other Protein Structure Databases and Table 6.1

The databases listed in Table 6.1 are not strictly primary protein structure databases. The table was expanded to include similar databases. What they have in common is that they support a full, or nearly full, complement of available protein structures.

Table 6.1. Primary Protein Structure Databases and Similar Sites

#	Database Name	Description	URL
Primary Protein Structure Databases			
1.	Protein Data Bank (PDB)	protein structures	www.rcsb.org/pdb/
2.	EBI-Macromolecular Structure Database (MSD)	protein structures	http://msd.ebi.ac.uk
Similar Protein Structure Databases			
3.	IMB Jena Image Library	protein, nucleic acid structures	www.imb-jena.de/IMAGE.html
4.	SWISS-3DIMAGE	protein structures	http://expasy.org/sw3d/
5.	Kinemages	protein structures	www.prosci.uci.edu/Kinemage/
6.	SWISS MODEL Repository	theoretical 3D models	www.expasy.org
7.	ModBase	theoretical 3D models	http://pipe.rockefeller.edu/ modbase-cgi/index.cgi
Other Image Databases Similar to Primary Protein Structure Databases			
9.	Protein Quaternary Structure Server (PQS)	tertiary protein structure assemblies	http://pqs.ebi.ac.uk/
10.	BioImage	microscopic images	www.bioimage.org
11.	WebMolecules	3D images of all molecule types	www.webmolecules.com

Secondary Protein Structure Databases

Secondary databases are filtered or altered versions of a primary database. The primary database is the PDB, the place where a researcher submits structure data to make it available through the web and, consequently, it is a complete set of all protein structure data. Secondary protein structure databases take data from the PDB and present it in one of three ways: as a subset of the PDB (filtered structure data sets); as a complete set of the data in the PDB (with data organized into related groups); or as a complete set of the data. When protein structure databases take a complete set of data, extra information is added to each protein file, specifically information regarding secondary structure—the secondary structure databases.

Structure Classification Databases

The classification of proteins by structure is in the same vain as the classification of proteins by sequence because the goal is to organize proteins based on similarity. The difference is that, though structure classification includes an initial sequence similarity grouping, the final assignment of proteins into families depends on structural similarity. Protein structures are classified based on their secondary structure composition (alpha helices and beta sheets) and the arrangement of the secondary structure (folding pattern creating the tertiary structure). Because different sequences can generate the same structures, there are less protein families based on structural similarity than there are for those based on sequence similarity.

Protein structure classification databases are useful because structural similarities may infer functional similarities, more so than sequence similarities. Classification databases are also useful for evolutionary studies as you can compare alterations in homologous structures.

The databases contain the same structures that the PDB contains, as evidenced by the fact that you can search them with a PDB ID. They are different because their files are organized into related sets (e.g. protein families), whereas with the PDB all the files are placed

Table 6.2. Secondary Protein Structure Databases

#	Database Name	Description	URL
Classification Databases			
1.	HOMSTRAD (+ PLUS)	homologous families	www-cryst.bioc.cam.ac.uk/data/align/
2.	FSSP	homologous families	www2.ebi.ac.uk/dali/
3.	CAMPASS	homologous families	www-cryst.bioc.cam.ac.uk/~campass/
4.	SCOP	hierarchy	http://scop.mrc-lmb.cam.ac.uk/scop/
5.	CATH	hierarchy	www.biochem.ucl.ac.uk/bsm/cath/
6.	PDB-REPRDB	representative chains of protein families	www.rwcp.or.jp/papia/
Filtered Structure Data Sets			
7.	BMCD	crystallography	wwwbmcd.nist.gov:8080/bmcd/bmcd.html
8.	BioMagResBank	NMR	www.bmrb.wisc.edu/Welcome.html
9.	Molecular Modeling DB	experimental data	www.ncbi.nlm.nih.gov/Structure/MMDB/mmdb.shtml
10.	CulledPDB	unique side chains	www.fccc.edu/research/labs/dunbrack/culledpdb.html
11.	PDBOBS	obsolete PDB entries	http://pdbobs.sdsc.edu/PDBObs.cgi
12.	Enzymes Structures DB	enzymes	www.biochem.ucl.ac.uk/bsm/enzymes/
13.	ABG	antibodies	www.ibt.unam.mx/vir/structure/structures.html
14.	3D_ali	published alignments	www.embl-heidelberg.de/argos/ali/ali.html
15.	PDBREPORT	structural errors	www.cmbi.kun.nl/gv/pdbreport/
Secondary Structure Databases			
16.	DSSP	secondary structure	www.cmbi.kun.nl/swift/dssp/

individually. Some of the databases allow you to view the structural alignments, the set of 3-D structures superimposed on one another. You can also view the sequence alignments that result from aligning by structure. The URLs of these databases are found in Table 6.2.

Homologous Families: HOMSTRAD, HOMSTRAD PLUS, CAMPASS, FSSP

Protein structures are classified either into homologous families or hierarchically. HOMSTRAD, HOMSTRAD PLUS, CAMPASS, and FSSP provide homologous family organization, isolated groupings (families) of evolutionarily related (homologous) proteins. HOMSTRAD's, HOMologous STRucture Alignment Database, classification scheme takes input from several protein sequence and structure classification databases and performs their own sequence similarity search in initial homologous family assignment. They do a significant function and structure residue alignment for a final adjustment of protein structures into homologous families. It is the manual editing of families in the last step of classification that makes this a high quality database. Manual editing takes longer than automated systems but provides a more accurate analysis. HOMSTRAD PLUS provides extra sequence family information to the HOMSTRAD structures. FSSP takes a fully automated approach to homologous family classification.

Hierarchical Classification: SCOP, CATH

Hierarchical organizations mean all proteins in the database are related to one another. Proteins are classified beyond the family level. For example, individual families are grouped into superfamilies. The Structural Classification of Proteins (SCOP) database's classification scheme includes a manual inspection of structures that is aided by automated approaches. Evolutionary relatedness (homology) is taken into account. They define proteins in a *Family* as having a clear evolutionary relationship due to obvious sequence similarities;

Superfamilies have a probable common evolutionary origin showing less sequence similarity than in a *Family*, but structural and functional similarities confirm the relationship; the most inclusive level in the hierarchical scheme is at the Fold level where "major structural similarity", secondary structures with the same arrangement and similar topology, defines protein relatedness.

CATH has a similar approach in that it places a heavy emphasis on sequence similarity, but when sequence comparisons are inconclusive, structural data takes precedence. CATH aims to increase the speed and sensitivity of its classification.

Filtered Data Sets

The Filtered Data Set databases contain a definable subset of structure data from the PDB. For example, the BMCD contains structure entries for crystallography determined structures only whereas CulledPDB contains unique, high quality side chains from the proteins in PDB. A list of databases falling in this category with a brief description of their PDB subset is found in Table 6.2.

The Molecular Modeling DataBase organizes and validates experimentally elucidated structures from the PDB. It was designed to serve as a resource for homology modeling. It reorganizes and validates the PDB data in its database. Other databases include only published alignments of protein structures (3d_ali), antibodies (ABG), enzymes (Enzyme Structures), structural errors in PDB files (PDBREPORT), and PDB entries that are now obsolete (PDBOBS).

Secondary Structure Databases: DSSP

The secondary structure database DSSP describes the secondary structure of a protein in the PDB. 3-D structures, the basis of all the data described above, represent proteins at the tertiary or quaternary folding state, the state at which a protein is functional. A secondary structure, the folding state that is associated with beta sheets and alpha helices, describes the structural components of a protein.

The PDB nearly always has information on secondary structure but its source is not disclosed. The PDB does not say who supplied the data nor the software used to determine it. The DSSP creates a standard in an otherwise subjective field.

Secondary structure annotation is presented by identifying the residues of the representative protein sequence that correspond to secondary structure features. There are tools described in Chapter 11, Proteomics Tools, which predict residues of an amino acid sequence involved in secondary structure. DSSP is different in that it uses 3-D structure coordinates as its data source.

Nucleic Acid Structure Databases

Most nucleic acid work involves looking at the sequence, but these molecules, DNA and RNA, do not always look the same structurally. Nucleic acid structures are not nearly as prevalent as protein structures, but their data has found a place on the web.

Nucleic Acid Structure Types

DNA is recognized simply as a double-stranded helix, which does not fully describe its structural capability. DNA has three conformations named A-DNA, B-DNA, and Z-DNA. There are also examples of quadruple-stranded DNA. Sometimes bases at the end of the helices are "flipped out" or somewhere within the helix a "bulge" may occur on one of the two strands. Nucleotide mismatches, such as a cytosine (C) bonding with an adenine (A), are another structural variation found in DNA.

RNA, a single-stranded molecule upon transcription, folds into a variety of structures including molecules that catalyze reactions (ribozymes), t-RNAs, r-RNAs, and RNA double helices. RNA structures from virus' and phages have also been determined.

A nucleic acid-protein complex is another type of structure to be found in a nucleic acid structure database. An example of this would be a transcription factor (protein) binding to DNA.

NDB (& PDB)

The NDB, Nucleic acids DataBase, and the PDB are closely associated. The PDB contains nearly all of the structures found in the NDB along with protein structures. The NDB focuses on nucleotide structures determined by X-ray crystallography but not structures determined by NMR (Nuclear Magnetic Resonance). It does provide a search of the PDB that covers the NMR structures.

NDB (http://ndbserver.rutgers.edu/NDB/) is an important database because it focuses on small data sets and is able to maintain itself as a thorough resource for nucleic acid structures. It provides plenty of background information and tutorials. Its small data set makes it easier to search. NDB takes advantage of this with their *Atlas of Nucleic Acid Containing Structures*.

The *Atlas* separates all the data of the database into categories. The categories include all structure types mentioned above. Clicking on any of these category names displays a full list of nucleotide structure files of that type found in the NDB. These lists generally contain 5 to 25 structures, making them easy to sift through.

The NDB provides basic text searches and downloading capabilities (a download of the whole database is available). The data itself is stored in the same file format as the PDB (PDB, mmCIF) and is displayed on the web interface very clearly.

3-D and 2-D images of the structures can be found at the bottom of a file. Static and interactive viewing is provided. The downloadable Rasmol viewer allows interactive viewing (described in Chapter 11, Proteomics Tools).

RNABase

The RNABase (www.rnabase.org) contains structural data for RNA only. On a daily basis, it automatically pulls structure files from both the NDB and the PDB, so it has a full-complement of all RNA structures available (about 500). It has yet to incorporate a way to submit data to it directly, so all RNA structure data should be submitted to the NDB or PDB.

Data is retrieved in the same manner with this database as with NDB. The small size of it allows for all the data to be categorized into RNA structure types (i.e., tRNA, rRNA, various complexes, hybrids, viral and phage RNA). The web interface is also very similar to the NDB. The obvious difference is that images are offered indirectly. A user must click on the NDB or PDB link (the database the data originally came from) to view the image.

IMB Jena Image Library and others

The IMB Jena Image Library of Biological Macromolecules (the URL can be found in Table 6.1) is considered a secondary resource of image data, though it provides access to all data deposited at the PDB and NDB. Its focus is in the visualization and analysis of 3-D structures. The data is arranged into a variety of categories, all found on the home page. Basic search tools are also provided and are worth examining. The data found in IMB is the same as that found in the PDB or NDB.

There are other databases with structural data, but they are more oriented towards sequence data. Most are small databases focused on

RNA (large subunit RNA, small subunit RNA, viral RNA, etc.), such as the Comparative RNA Web Site. Some of these databases are listed in Chapter 4, Secondary Sequence Databases.

Further Reading

Berman, H.M., Westbrook, J., Feng, Z., Gilliland, G., Bhat, T.N., Weissig, H., Shindyalov, I.N., and Bourne, P.E. 2000. The Protein Data Bank. Nucleic Acids Res. 28: 235-242.

Garavelli, J.S. 2000. The RESID Database of protein structure modifications: 2000 update. Nucleic Acids Res. 28: 209-211.

Holm, L., Sander, C. 1998. Touring protein fold space with Dali/FSSP. Nucleic Acids Res. 1: 316-319.

Lo Conte, L., Ailey, B., Hubbard, T.J.P., Brenner, S.E., Murzin, A.G.and Chothia, C. 2000. SCOP: a Structural Classification of Proteins database. Nucleic Acids Res. 28: 257-259.

Mizuguchi, K., Deane, C.M., Blundell, T.L., and Overington, J.P. 1998. HOMSTRAD: a database of protein structure alignments for homologous families.Protein Sci. 11: 2469-2471.

Noguchi, T., Matsuda, H., and Akiyama, Y. 2001. PDB-REPRDB: a database of representative protein chains from the Protein Data Bank (PDB).Nucleic Acids Res. 29(1): 219-220.

Pearl, F.M.G., Lee, D., Bray, J.E., Sillitoe, I., Todd, A.E., Harrison, A.P., Thornton, J.M., and Orengo, C.A. 2000. Assigning genomic sequences to CATH. Nucleic Acids Res. 28: 277-282.

Reichert, J., Jabs, A., Slickers, P., and Sühnel, J. 2000. The IMB Jena Image Library of Biological Macromolecules. Nucleic Acids Res. 28: 246-249.

Sowdhamini, R., Burke, D.F., Huang, J.F., Mizuguchi, K., Nagarajaram, H.A., Srinivasan, N., Steward, R.E., and Blundell, T.L. 1998. CAMPASS: a database of structurally aligned protein superfamilies. Structure. 6(9): 1087-1094.

Wang, Y., Wang, Y., Addess, K.J., Geer, L., Madej, T., Marchler-Bauer, A., Zimmerman, D., and Bryant, S.H. 2000. MMDB: 3D structure data in Entrez. Nucleic Acids Res. 28: 243-245.

Chapter 7

Gene Function Databases: Enzymes, Interactions, Expression, and Pathways

---

Contents

From: *Genomes and Databases on the Internet: A Practical Guide to Functions and Applications*
ISBN 1-898486-31-X © 2002 Horizon Scientific Press, Wymondham, UK.

Abstract

Identifying genes through sequencing efforts does not solve functional secrets. Knowing the function of a gene involves knowing where it fits in the complex system of interactions inside cells: what molecules does it interact with, what are the products of the interactions, etc. Getting these answers involves many steps, including determining the function of individual genes by looking at the other molecules they interact with and piecing together individual interactions into a web of relations. The databases in this chapter fill voids at these steps. to create a tangible model of how a cell operates. Molecular pathway databases (KEGG, WIT, etc.) present tangible models incorporating multiple interactions; databases of interactions (DIP, EMP, BRENDA, ENZYME, etc.) provide the pieces; and gene expression studies (Stanford Microarray Database, etc.) lead to clues in satisfying both quests. This chapter explains the databases involved in this effort.

Introduction

Sequence, structure, and classification databases are compiling vast amounts of data on individual genes. The genes are described with much detail, culminating in a functional description. This one-to-one relationship between a gene and its function is meshed with other genes of known function in trying to explain the complex system of interaction that takes place within cells. The databases in this chapter aim to put together individual pieces of data in defining how a cell operates.

There are different pathways of interaction in a cell. The three most common pathway types are metabolic, regulatory, and signal transduction. Metabolic pathways are chemical changes where energy is supplied for various processes and new materials are generated. Essentially, all other pathways are a subset of metabolic pathways. Regulatory pathways determine what processes are active and their level of activity. Signal transduction pathways, a subset of regulatory pathways, communicate information from extracellular stimuli to the nucleus of a cell through a cascade of interaction. These pathways involve many molecules in a step-by-step series of interactions.

Pathways can be dissected into individual events: the interaction of one molecule to another, usually an enzymatic interaction or a protein-protein interaction. Enzyme databases and interaction databases are described separately in this chapter. In addition, gene expression databases based on microarray data are also included because microarrays have tremendous potential in deciphering significant amounts of information regarding gene function and interaction. Finally, the pathway databases are reviewed.

Figure 7.1 shows an example of a pathway and associated data retrieved from the TRANSPATH database. TRANSPATH is discussed separately later in the chapter.

Enzyme Databases

In general, enzymes are proteins that catalyze reactions. Databases that focus on enzymes look at molecules individually. They provide hints about a pathway by incorporating information on the molecules involved in the reaction. Often, those molecules in the reaction are proteins. These databases then act as the foundation of pathway databases by presenting the individual segments that make up a pathway.

Enzyme databases are organized according to Enzyme Commission (EC) numbers. These numbers are made official through the publication *Enzyme Nomenclature*, which includes all enzymes whose existence has been confirmed by the scientific community. The naming

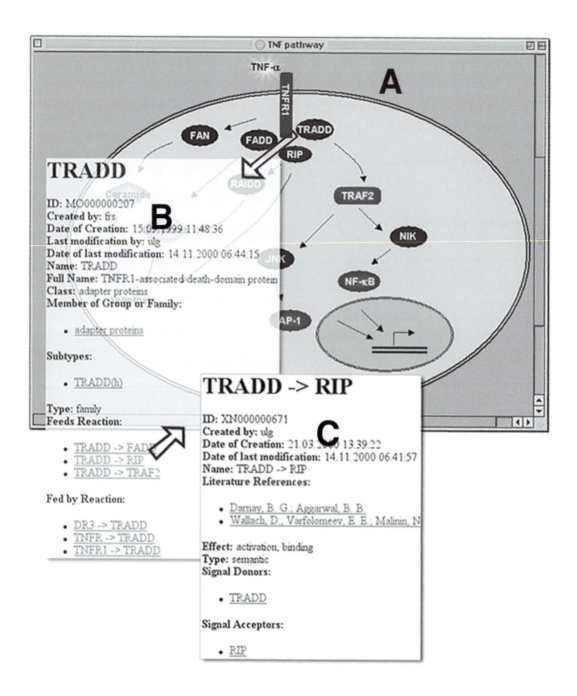

Figure 7.1. This is an example of the type of data generated by databases in this chapter. The images for this figure were obtained from the TRANSPATH database. Clicking on the pathway graphic (A) retrieved text-based data for a molecule (B). The molecule page links up to data describing the reactions it is involved in (C).

(and numbering) of enzymes has been monitored since the first publication of *Enzyme Nomenclature* in 1961. The sixth and latest edition of *Enzyme Nomenclature*, published in 1992, can be found on the web. The web version, served by the International Union of Biochemistry and Molecular Biology (IUBMB), includes supplements and changes to the print edition of 1992. These additions will be part of the seventh publication.

Enzyme Nomenclature (on the web) and EC Numbers

Enzyme Nomenclature (see Table 7.1 for the URL) not only provides the official names of enzymes. It also provides information on the reaction of the enzyme it catalyzes, alternate names of the enzyme, comments, links to other databases, and literature references.

The web version is a fully functioning database of information as it provides a means to mine the data: a search feature and a sorted list of enzymes. The list is sorted according to EC numbers. An EC or Enzyme Commission number looks like this: "EC 1.1.1.1". There are six enzyme classes referring to Oxidoreductases (EC 1.#.#.#), Transferases (EC 2.#.#.#), Hydrolases (EC 3.#.#.#), Lyases (EC 4.#.#.#),, Isomerases (EC 5.#.#.#), and Ligases (EC 6.#.#.#). For example:

6.#.#.# = Ligases
6.3.#.# = Ligases forming carbon-nitrogen bonds
6.3.3.# = Cyclo-ligases (of those ligases forming carbon-nitrogen bonds)
6.3.3.1, 6.3.3.2, and 6.3.3.3 = the three cyclo-ligases

EC numbers are seen as links in almost every database we describe in this book. The links act as cross-references to other databases and are usually found at the bottom of a data file.

Table 7.1. Enzyme Databases

#	Database	URL
1.	Enzyme Nomenclature	www.chem.qmw.ac.uk/iubmb/enzyme/
2.	ENZYME	www.expasy.org/enzyme/
3.	LIGAND	www.genome.ad.jp/dbget/ligand.html
4.	BRENDA	www.brenda.uni-koeln.de/

ENZYME, LIGAND, and BRENDA

ENZYME, LIGAND, and BRENDA are alternate databases whose focus are enzymes. They incorporate more information than Enzyme Nomenclature in an effort to bring the enzymatic data (mostly chemical) together with biological data. The added data includes more functional information, structural information, and even experimental techniques to isolate enzymes (BRENDA). Due to added biological data, the query systems are more extensive. The following paragraphs offer a brief description of alternate databases.

ENZYME is part of the ExPASy system, which includes SWISS-PROT and PROSITE, databases for protein sequence and classification, respectively. This allows ENZYME entries to reflect data found in SWISS-PROT and PROSITE.

LIGAND is closely affiliated with KEGG, the metabolic pathway database described later in this chapter. Ligand presents individual reactions from KEGG pathways.

BRENDA is a very thorough resource of enzymatic data. It gathers data from publications, which explains why it has information regarding the experimental isolation of enzymes. It is a great resource for those interested in any lab work involving enzymes.

Interaction Databases

Databases of molecular interactions, most commonly of protein-protein interactions, are similar to enzyme databases because they act as a foundation for the pathway databases. The interactions they describe are pieces of a larger picture. While the enzyme databases refer to the changing of a compound from one form to another with the aid of an enzyme, interaction databases, or more directly protein-protein interaction databases, refer to two proteins that bind to each other for functional purposes. These functional purposes can be the same as those described for enzymes, however they can also be a step in a regulatory or signal transduction pathway.

Table 7.2 lists the names of the databases and their URLs.

BRITE

BRITE, Biomolecular Relations in Information Transmission and Expression, is a database closely associated with KEGG and LIGAND. Its focus is on protein relationships, which includes information originating from KEGG pathways, experimental research, sequence similarity studies, and gene expression studies.

DIP

DIP, Database of Interacting Proteins, contains information on protein-protein interactions only. The data on an interaction includes general information on the proteins involved (name and links to SWISS-PROT, PIR, GenPept). Data for individual proteins also includes the protein domain involved in the interaction, the protein's classification, and a list of all the other proteins that it interacts with. A Java applet provides a graphical display of the interactions.

Table 7.2. Interaction Databases

#	Database	URL
1.	BRITE	www.genome.ad.jp/brite/
2.	DIP	http://dip.doe-mbi.ucla.edu/
3.	BIND	www.bind.ca/
4.	ProNet	http://pronet.doubletwist.com/

BIND

BIND, Biomolecular Interaction Network Database, covers molecule-molecule interactions of proteins, nucleic acids and small molecules, nearly 6,000 in all. The interactions range from small molecule biochemistry to signal transduction, including chemical reactions as well as conformational changes. BIND covers pathways (6 of them) and molecular complexes (51). A Java applet allows for a graphical display of the interactions.

ProNet

ProNet is a protein-protein interaction database. The immediate goal is to get every published interaction into the database. From there, it will remain up to date through direct submissions from the scientific community, continued perusal of the published literature, and from their own high-throughput experimentation designed to elucidate protein-protein interactions. Currently, the database only contains human protein interactions, but it will expand to other organisms in the future. ProNet's Java display is the easiest to use and the most interesting of the interaction displays.

Gene Expression (Microarrays) Databases

Genomic DNA sequencing, whose role is to depict the genes organisms, is an important first step to understanding the molecular

Table 7.3. Microarray Databases and Related Resources

# Site Name	URL
1. Stanford Microarray Database (SMD)	http://genome-www4.stanford.edu/cgi-bin/SMD/login.pl
2. ArrayExpress	http://www.ebi.ac.uk/arrayexpress/index.html
3. GeneX at NCGR*	http://www.ncgr.org/research/genex/
4. Microarrays.org	http://www.microarrays.org
5. Microarray Gene Expression Group (MGED)*	http://www.mged.org
6. Gene Expression Markup Language (GEML)*	http://www.geml.org
7. NetGenics*	http://www.netgenics.com
8. Object Management Group (OMG)*	http://www.omg.org

Notes: * sites where you can find information on the data standardization process.

activity of life. Organisms do not use every gene all the time and at equal rates. A Microarray is a relatively new technology that allows you to determine which genes are expressed and at what levels. In the past, expression studies previously examined genes individually, but microarrays allow you to examine the expression of thousands of genes simultaneously. This is done by spotting sequences that are representative of different genes onto a substrate, generally a glass slide, and probing the spotted array with a labeled probe. The results are an expression value for each gene spotted.

With the tremendous amount of sequence data available, the scientific community has many genes of unknown function. Microarrays help determine the function of a gene by way of expression data.

Similar to the GenBank, EMBL, and DDBJ nucleotide sequence databases, a system for collaboration and data standardization among the public microarray databases is in progress. A universal format for annotation and data representation is of great importance for any public database because many different people submit data and an even greater number view and attempt to use the data. Due to the large quantity of microarray data, this is of even greater importance.

The sites listed in Table 7.3 contain databases that are both fully functioning and developing. Sites are listed with information on the data standardization process.

Pathway Databases

Pathway databases compile data from a variety of sources in presenting a holistic view of cellular processes in the form of pathways. Among others, these include metabolic pathways, regulatory pathways, and signal transduction pathways. Table 7.4 lists the databases and their websites.

KEGG

KEGG, Kyoto Encyclopedia of Genes and Genomes, in their own words, "is a knowledge base for systematic analysis of gene functions in terms of the networks of genes and molecules." KEGG integrates data from chemical and biological sources of all types to explain the complex nature of cellular operations. The chemical and biological sources of data include other pathway databases such as WIT, the primary sequence and structure databases, GenBank, SWISS-PROT, PDB, as well as LIGAND, BRENDA, and PROSITE.

The wealth of information KEGG has to offer is accessed through interactive graphical representations of pathways and genome maps, tables and catalogs of data, and computational (analysis) tools.

The pathways, organized into the PATHWAY database, are a collection of metabolic, regulatory, and molecular assembly pathways. PATHWAY contains all known metabolic pathways, such as the TCA cycle. Currently, it is increasing its collection of regulatory pathways (i.e., MAPK signaling pathway) and molecular assembly pathways (i.e., ribosome assembly). The pathways are interactive because you can click on different sections of the diagram, such as a box representing an enzyme, to access more detailed information. There are *Reference* pathways and organism-specific pathways. The *Reference* pathway includes all chemically feasible components, whereas the organism specific pathways highlight those components, such as an enzyme in a metabolic pathway, that are known for that organism (Figure 7.2).

The genome maps are an interactive graphical representation of completely sequenced genomes. PATHWAY is linked to the genome maps enabling quick retrieval of a pathway for a gene of interest found on the map (Figure 7.2). KEGG also supports comparative genome maps that take two genomes and compare their genetic makeup, leading to discoveries of similar genes between genome. The genome's sequence comes from the completed genomes section of GenBank. The textual information is found in the GENES database.

The GENES database is organized into complete genomes whose genes are fully annotated. An entry for a gene includes its nucleotide and protein sequence, functional data, pathway classification, and chromosomal position. The genes are catalogued according to their presence in a pathway. The GENES database is searchable through the DBGET/LinkDB system (see Chapter 10).

Other sources of information are the ortholog group tables and the molecular catalogs. The ortholog groups consist of genes organized into groups of functional similarity. The molecular catalogs provide classification of all molecular types—proteins, RNA, other biological macromolecules, chemical substances and molecular assemblies— based on structural and functional data.

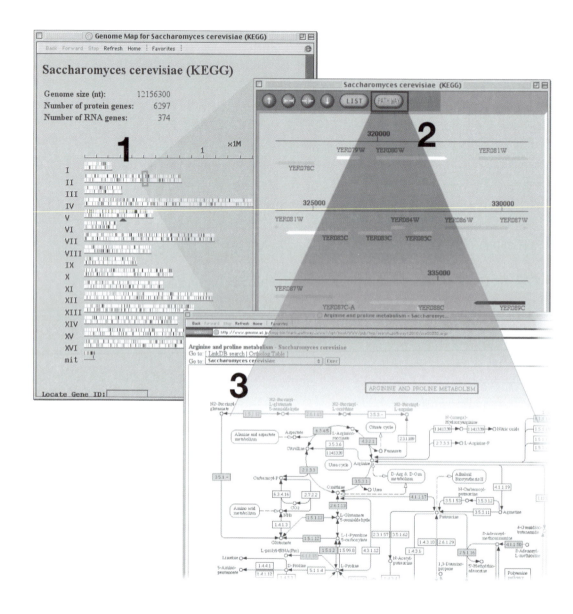

Figure 7.2. KEGG offers very helpful graphical displays of its data. An interactive display of *S. cerevisiae* chromosome architecture (1). Zooming in on a section of Chromosome II (2). Pathway representation of a gene from Chromosome II (3).

The computational tools KEGG offers relate to searching and analyzing the pathway maps and genome maps. For example, a coloring tool allows you to color code a pathway based on microarray gene expression data. Other tools allow you to reconstruct a pathway by finding alternate paths in a network of interactions.

KEGG's *Table of Contents* page clearly organizes the resources it offers and the DBGET/LinkDB system is a very useful tool in querying the many facets of KEGG.

WIT

WIT's, "What Is There?", objectives are very similar to those of KEGG. Data and ideas are shared collaboratively between the two databases. This leads to a higher standard of information, but also similar information. WIT represents its data differently. Similar to other sets of databases that work collaboratively (i.e., EMBL vs. GenBank vs. DDBJ), it is best to use both to see which is best for you.

WIT is similar to KEGG because it bases its approach on metabolic pathways. It generates a metabolic reconstruction of individual organisms from all known metabolic pathways. A metabolic reconstruction is a model of an organism's metabolism. WIT determines what pathways are present in an organism and which genes implement the functional roles. The results of this work are best understood by describing the types of tables used to display data (interactive diagrams of metabolic pathways are also available). There is a reference table, comparable to KEGG's *Reference* diagrams, which lists the generic pathways and the functional roles of genes in the pathways. There are also two organism specific tables: one listing the pathways of an organism and the other listing the genes of the organism, including the pathway the gene is in, and the functional role of the gene in the pathway.

In generating a metabolic reconstruction for an organism, WIT annotates open reading frames (ORFs), or genes, from genome

Table 7.4. Pathway Databases

#	Database	URL
1.	KEGG	www.genome.ad.jp/kegg/
2.	WIT	http://wit.mcs.anl.gov/WIT2/
3.	EMP	http://emp.mcs.anl.gov/
4.	TRANSFAC System	http://www.gene-regulation.de/
5.	CSNDB	http://geo.nihs.go.jp/csndb/

sequencing projects. Their annotation is more reliable than most—usually annotation is done with one gene in mind—because functional roles among genes must coordinate to fit into a metabolic reconstruction. WIT works on both fully and partially sequenced genomes (39 in all as of this writing).

WIT also provides information on orthologs and operons. The "Ortholog" data refers to a group of genes that are similar to each other. The group contains only one gene from each organism. The genes, by definition of being in an ortholog, are supposed to carry the same functional purpose, but this is not always the case. Because of this, WIT has divided its ortholgous groups into three sections based on their level of functional similarity. WIT explains the inconsistencies in ortholog representation quite well.

Operons, another term WIT points out as being misrepresented all too often, refer to a group of genes on a chromosome. WIT matches operons among organisms, just as it matches individual genes in creating orthologous groups.

WIT provides a toolbar accessible on every page with direct access to a *General Search*, *View Model* (diagrams and other data of metabolic pathways), *Similarity Search*, *Ortholog Clusters*, *Operons*, *Query Pathways*, and *Query Enzymes* among others.

EMP

The Enzymes and Metabolic Pathways database (EMP) covers all aspects of enzymology and metabolism found in the published literature: biochemical genetics, enzyme regulation, enzyme structure, equilibrium and thermodynamics, physical chemistry, immunochemistry, etc. EMP adds nearly 8,000 entries a year and is currently approaching 30,000. The metabolism data includes over 3,000 diagrams of metabolic pathways. EMP, formerly know as the Metabolic Pathway Database, collaborates with WIT; therefore, their diagrams are the same.

TRANSFAC System

The TRANSFAC System is a set of databases dedicated to the study of transcription factors, proteins that are the primary regulators of gene expression. Transcription factors bind to DNA in initiating the transcription of a gene. Genes are often transcribed in response to what is taking place outside of the cell. Signal transduction pathways relay information from extracellular stimuli to the nucleus of a cell, culminating in a transcription factor response. The pathways outlined in TRANSPATH, one of six databases in the TRANSFAC System, focus on signal transduction pathways. The data consists of diagrams of pathways that include the individual steps. The individual steps can be linked to more information on the reaction itself or the molecules involved in it. See Figure 7.1 for an example of what one might find in the TRANSPATH database.

CSNDB

CSNDB is the Cell Signaling Networks DataBase. It focuses on the signal transduction pathways in human cells only. Pathways are elucidated through the compilation of various biological data corresponding to cell signaling. This data can be found in the literature.

CSNDB offers diagrams of the pathways, which can be obtained by browsing or searching the database. CSNDB is linked to the TRANSFAC database.

Further Reading

Bader, G.D. *et al.* 2001. BIND—The Biomolecular Interaction Network Database. Nucleic Acids Res. 29: 242-245.

Bairoch, A. 2000. The ENZYME database in 2000. Nucleic Acids Res. 28: 304-305.

Goto, S., Nishioka, T., and Kanehisa, M. 2000. LIGAND: chemical database of enzyme reactions. Nucleic Acids Res. 28: 380-382.

Kanehisa, M., and Goto, S. 2000. KEGG: Kyoto Encyclopedia of Genes and Genomes. Nucleic Acids Res. 28: 27-30.

Overbeek, R., Larsen, N., Pusch, G.D., D'Souza, M., Jr, Kyrpides, N., Fonstein, M., Maltsev, N., and Selkov, E. 2000. WIT: integrated system for high-throughput genome sequence analysis and metabolic reconstruction. Nucleic Acids Res. 28: 123-125.

Selkov, E. Jr, Grechkin, Y., Mikhailova, N., and Selkov E. 1998. MPW: the Metabolic Pathways Database. Nucleic Acids Res. Jan 1; 26(1):43-5.

Takai-Igarashi, T., and Kaminuma, T. 1999. A pathway finding system for the cell signaling networks database. Silico Biol. 1(3):129-46.

Wingender, E. *et al.* 2001. The TRANSFAC system on gene expression regulation. Nucleic Acids Res. 29: 281-283.

Xenarios, I. *et al.* 2001. DIP: The Database of Interacting Proteins: 2001 update. Nucleic Acids Res. 29: 239-241.

Part II

Genomic Research

Chapter 8

Genomics Centers

Contents

Abstract

Multiple genome analysis requires computing power and manpower beyond a typical lab's capabilities. Genomic centers have the resources to construct secondary genome databases for more than one genome. These genome libraries showcase the progress of the sequencing efforts and are another entry point into the primary sequence data. Genomic centers make use of genome map viewers and text-based browsers to view genome data comparatively.

From: *Genomes and Databases on the Internet: A Practical Guide to Functions and Applications*
ISBN 1-898486-31-X © 2002 Horizon Scientific Press, Wymondham, UK.

Introduction

Genomes are sequenced at various sequencing facilities around the world. Those facilities also annotate the sequence they generate by identifying where the genes are in the sequence. The gene sequences are then dispersed to the scientific community through databases. The primary nucleotide sequence databases of Chapter 1 (EMBL, GenBank and DDBJ) present this sequence data as individual entries, whereas a genome center, the focus of this chapter, compiles the sequence data as it relates to genomes. Genome centers are different from genome databases because they offer access to sequence data of multiple organisms and genome databases focus on one organism. Four genome centers are described below. URLs for these centers are found in Table 8.1

Entrez Genomes

The Entrez genomes page at NCBI is the collection of genome resources at NCBI. Here you can browse a subset of the approximately 800 genomes available in the GenBank database. The Entrez genomes page produces views of genomes, complete chromosomes, mapped sequences, and integrated physical and genetic maps. The map viewer is a graphical interface to the genome projects and provides a number of summaries all within a linked representation of the genome or chromosome. Clicking on a region of interest will zoom and center the chromosome image on the area of interest. Eventually the display will give way to the GenBank record for a particular gene within the Entrez search system. Querying organisms with smaller genomes leads to the genome record in Entrez that can be downloaded or queried further.

EBI Genomes

The EBI Genomes' text-based browser is much less visual than the genome resources at NCBI. EBI Genomes provides a table of genomes

Table 8.1. Genomics Centers

#	Site Name	URL
1.	Entrez genomes	www.ncbi.nlm.nih.gov/Entrez/Genome/org.html
2.	EBI genomes	www.ebi.ac.uk/genomes/
3.	TIGR Genomes	www.tigr.org/tdb/
4.	NCGR	www.ncgr.org/

and links to either their chromosomes for eukaryotic organisms or to genome segments for smaller genomes. Much like the Entrez genomes, this leads to an EMBL sequence report page. The sequence report page is a FASTA formatted page of sequence data except for the human genome project an EnsEMBL reports page is generated. Unlink Entrez Genomes, this is the end of the query capabilities for EBI Genomes. Additional information about the genome must be investigated with the other search applications on the EBI website.

TIGR Genomes

TIGR has been a key source for microbial genome sequences and makes much of its sequence data available online. TIGR Genomes provides a range of tools for the investigation of their genome projects. The *Arabidopsis thaliana* genome project at TIGR has a map viewer much like those of Entrez Genomes that profiles the sequence and annotation efforts of the genome project. Each element in the map viewer is linked to a database record and can be clicked for additional information. TIGR has also collected information on microbial genomes on their Comprehensive Microbial Resource (CMR) web page (see Chapter 9 for more information). The other databases are accessible through sequence or keyword search pages.

NCGR

The National Center for Genomic Resources has a myriad of genomics tools that includes microarray analysis software. NCGR also houses genome databases for the model legume *Medicago* and the potato famine fungus *Phytopthora*. NCGR collaborates with the Carnegie Institution at Stanford in maintaining The Arabidopsis Information Resource (TAIR). See Chapter 9 for details on TAIR.

Chapter 9

Genomes

Contents

From: *Genomes and Databases on the Internet: A Practical Guide to Functions and Applications*
ISBN 1-898486-31-X © 2002 Horizon Scientific Press, Wymondham, UK.

Plant Genome Databases
    UK CropNet / ARS GDR
    TIGR Plant Genomes
    Arabidopsis
        AGI – TAIR
        MATDB
        AGR Arabidopsis Genome Resource (UK CropNet)
        TIGR's AtDB
    Banana Genomics
Insect Genome Databases
    Drosophila
        Flybase
        The Interactive Fly
        Flybrain
        FlyView
        FlyMove
    Mosquito
        Mosquito Genomics
Fungus Genome Databases
    Yeast
        Saccharomyces Genome Database (SGD)
        Yeast Proteome Database (YPD)
        MIPS Yeast Genome Database (MYGD)
Invertebrate Genome Databases
    C. elegans
        WormBase
        AceDB
        C. elegans AceDB browser at the Sanger Centre
        WormPep
    Dictyostelium
        Dictybase
Bacterial Genome Databases
    General Bacterial Genome Databases
        World Data Centre for Microorganisms (WDCM)
        Comprehensive Microbial Resource (CMR)
        Microbial Genome Database (MGDB)
        HOBACGEN: Homologous Bacterial Genes Database

Abstract

Genome sequencing projects are steadily finishing off the genome after genome. Currently there are over 800 genomes either in the process of being sequenced or already have been sequenced. There are enough complete genomes at present where many websites have been created in dedication to a specific organism. The dedication results in something far greater than a collection of nucleotide sequences from one organism. This chapter analyzes nine different genomic groups from human to plant to virus, pointing out the features that distinguish genome databases from collections of sequence.

Introduction

Primary sequence data is the raw information from the sequencing projects and published literature from around the world. Primary sequence databases act as the data distributing centers while secondary sequence databases repackage and add additional information. The secondary sequence databases in this chapter focus on the genome specific resources for a number of organisms from viruses and organelles to humans and plants. Within each of the secondary database categories there are multiple types of resources depending on the size of the genome, the type of organism, and the purpose of the sequencing efforts. For example, the genome resources for the human genome project are much different than those for bacterial or viral genome projects. The human genome project has developed computational tools for the annotation and assembly of large amounts of sequence data while bacterial and viral genome projects concentrate their efforts on homology searches by aligning the sequences of multiple bacterial genomes.

This chapter will profile a number of the resources for nine different areas of organism specific genome research. Multiple databases are reviewed for each of the following genomes:

1. Human
2. Vertebrates
3. Plants
4. Insect
5. Fungus
6. Invertebrate
7. Bacteria
8. Organelle
9. Virus

Human Genome Databases

The Human genome project led by Human Genome Initiative (HGI) is the largest undertaking of its kind. It has brought together a world of scientists and created the most nucleotide sequence data for a single organism to date. The sequencing of the human genome, completed in the spring of 2000 (years ahead of schedule) signified the unlocking of genetic information for computational investigation. To help support the sequencing effort, the members of the project and other computational biology labs have created a myriad of genome resources, each with a distinct role in the completion of the project. Most of the computational resources have been made available online via the Internet and provide valuable tools for human and nonhuman biologists alike. This section of the organism genome chapter profiles a handful of the many tools available for human genome analysis. The Online Mendelian Inheritance in Man (OMIM), a database of human genes and genetic disorders, is an excellent example of a genome project by-product that has become an established tool in its own right. This listing of gene descriptions has been used as the source of content on a number of human databases. Genome browsers are another type of tool used for the analysis of genome sequence. The Human Genome Database (HGD) and the UCSC Golden Path are two of the many genome browser tools that can be used to query the human genome sequence via search pages or graphical interfaces. As research on the human genome progresses, websites such as GeneCards focus research on web usability and data integration to bring the complex "labyrinth of data" to researchers in a friendly format. It should be noted that this list is not an exhaustive list of the resources available on the Internet, but it should point online researchers in the right direction for their area of interest and give a glimpse as to what is available online. URLs for the sites we mention are found in Table 9.1.

HUGO Gene Nomenclature Committee

The Human Genome Organisation (HUGO) consists of a panel of scientists who are responsible for gene name validation. One of the many administrative roles of HUGO is to ensure new gene names do not conflict with existing names and to check for compliance with the naming conventions used for the genome project. The committee has posted a protocol for how to devise new names for genes and literature aliases. Gene names are kept in the Human Gene Nomenclature Database: Genew3 database. Together the protocol and searchable database provide a valuable resource for human genome researchers.

Online Mendelian Inheritance in Man (OMIM)

Adding meta data or descriptions and references to genes is becoming more important as the genome project turns to assigning gene function to the raw sequence data. The OMIM database at NCBI profiles the genes and genetic disorders in humans. Each of the database entries is a summary of a gene products role in the human body with extensive links to related genomic information at NCBI. The database has been integrated into the Entrez toolbar which allows easy access to all of the available information at the website. The other entry points for the database include the OMIM Gene Map. This page lists the human genes organized by chromosome number, a search page where keywords can be searched against the database and the OMIM Morbid Map, a search page of the genetic disorders in man. This database is extensively referenced via hyperlinks in numerous other secondary databases for the human genome and serves as the primary text based portal for human genes.

Gene Atlas

The Gene Atlas website is a compilation of data from the human genome mapping efforts. This data has been collected from genome workshops, primary sequence databases and the published literature. The data is divided into three directories: GENATLAS/GEN, GENATLAS/LINK, and GENATLAS/REF, which correspond to gene data, the linkage of genes with disorders, and cross-reference of genes with their mutations, and disorders, respectively. The Gene Atlas database has a graphical interface of chromosomes that can be clicked through to browse the human genome. Gene Atlas is a unique interpretation of the human genome and a valuable comparative resource for genome biologists.

EnsEMBL

The focus of EnsEMBL is the sequence annotation and assembly of the human genome. There are many paths into the EnsEMBL sequence

Table 9.1. Human Genome Databases

#	Site Name	URL
1.	Human Genome Project	http://www.nhgri.nih.gov/HGP/
2.	HUGO	http://www.gene.ucl.ac.uk/hugo/
3.	OMIM	http://www.ncbi.nlm.nih.gov/Omim/
4.	EnsEMBL	http://www.ensembl.org/
5.	GDB	http://gdbwww.gdb.org/
6.	Gene Cards	http://bioinfo.weizmann.ac.il/cards/
7.	Allgenes.org	http://www.allgenes.org
8.	UCSC Golden Path	http://genome.ucsc.edu/

annotation efforts. The most obvious is the chromosome image map at the start page of the site. This linked image lets you start browsing the human genome at any one of the chromosomes and narrow your search to the genetic locus of interest. Once at the gene record page researchers may view the justification for the assignment of gene function or homology.

The EnsEMBL software package is a suite of software tools for assembling and tracking large scale sequencing products. The system downloads sequence from primary sources and subsequently assembles and stores it into the EnsEMBL database. Once a consensus sequence has been developed for an unknown string of sequence by both the database and researchers, the data is made data available via the web. This is not a trivial process and tracking the versions or builds of the genome has become less of a burden with the help of the EnsEMBL database. The database has been configured to holds genes, Single Nucleotide Polymorphisms, DNA Repeats and other Homologous Sequences. One of the interesting things about the project is the availability of the software, which allows customization of the database. The software can be used for any genome sequencing project and has been adapted by EBI to work with the mouse genome project. To use the software it must be downloaded to a local machine and installed. See the EnsEMBL database page for more information on system requirements and setup instructions.

Since EnsEMBL gets raw sequence from primary databases, it must annotate the genes. It does this via the GeneScan software, which searches the sequence for regions of DNA that matches either known gene domain patterns, sequence repeats or polymorphisms. The resulting sequence elements are kept and later compared to a database of known genes to help assign the function of the gene via comparative or homologous function assignments. Once a putative identity has been assigned to a gene, it is tagged with an EnsEMBL number and a version. If the annotation of the gene is changed or the sequence altered, the version is also changed; this assists in the tracking of gene annotation for such a widespread, collaborative effort.

The web interface can be used as a search engine for an OMIM disease entry, BLAST, or visually via the chromosome browser.

Genome Data Bank

The Genome Data Bank (GDB) is the sequence repository of the Human Genome Initiative. The initial focus of the project was to collect and annotate the mapping information for the sequencing centers. As the project has progressed the data bank has adapted and now provides information about the DNA sequence and the efforts to annotate the unknown genes. The database has been divided into three areas corresponding to the types of data objects represented in the database.

The Genomic Segment object is used by the database to describe genetic elements of all kinds. More specifically the genomic segments are regions of the genome ranging in size from one base to whole genes, contigs or sequence repeats. This is analogous to the features used in the gene bank record. (See Chapter 2)

The Map object is a common theme among genome browsers, the genome maps display a region of interest in the context of the map. The GDB map resources include contig maps, integrated homology maps, radiation hybrid maps, cytogenic maps and linkage maps. Each of these maps may be browsed via online graphical interfaces or text browsers at the GDB website.

The final object the Variation object keeps track of genetic variation within genomic sequence. At the time of this writing this includes 18,000 polymorphisms including allele information and links to the materials and methods used to derive the polymorphism.

GeneCards

GeneCards is one of the few databases to focus on the user interface. To assist online biologists the website has included questionnaires, a synonym database to resolve gene names, and a feedback form to keep the users involved in the development process. The gene records are also linked to various practical gene databases such as the Doctors Guide to the Internet and The Tumor Gene Database.

The Weizmann Institute in Israel developed GeneCards as a catalogue of genetic information that facilitates the searching of various human genetic databases. Another priority of the GeneCards project is the medical applications of the human genome data. The project aims to be one of the most useful available on the web by concentrating on the usability of the web interface and providing extensive links to other practical databases. They have created an easy-to-use interface to the "labyrinth of data" that has spread over the web. One of the tools they have devised is a query checker that will assist your search by suggesting a different spelling or a third party database that might be better suited to your needs. After entering the query, the list of possible matches is displayed with a brief description of the entry. The GeneCard record contains an informative summary of the gene of interest with descriptions and links to other relevant information.

Allgenes.org

Allgenes.org uses the murine - human genetic relationship to help interpret genetic data. The goal of the project is to compile a complete set of mouse and human genes into one publicly accessible database. While this is a common theme among biological databases, Allgenes.org employs all types of sequence data (gene description, phenotypes) using the controlled vocabularies and gene ontologies to structure sequence searches. To facilitate the homology search, Allgenes.org employs TIGRs cellular roles controlled vocabulary. This creates a common denominator for gene function by providing a clear set of words to describe the role of genes in the cell. A practical online application of this is a Javascript-enabled query page that helps to traverse the vocabulary hierarchy to find the right words to describe the gene or gene cluster of interest.

The database at Allgenes.org consists of human and mouse predicted genes using two methods: the Database Of Transcribed Sequences (DoTS) database developed at the In-house bioinformatics facility (U Penn) and the first pass gene finding software applications GRAIL-EXP and GENESCAN from the Oak Ridge National Laboratories.

There are three entry points into the Allgenes.org database. This first and most standard way for nucleotide-based databases is BLAST. The second is a submission-based search of the database in which you can sort the entries based on predefined query options. Finally, there is the Boolean query option that allows you to submit a uniquely defined query based on Boolean or True/False statements.

UCSC Golden Path

The first complete assembly of the human genome was achieved by a small group at the University of California at Santa Cruz led by Dr. Dave Haussler. This achievement was the result of the sequencing efforts of the International Human Genome Project and the Human Genome Mapping Consortium. The complete "working draft" assembly can be downloaded from the UCSC Human Genome site or browsed via the online genome browser. BLAT, a sequence search tool, works much like BLAST by submitting a sequence or set of sequences to the genome database and returning a series of alignments after the search has been completed. BLAT is different from BLAST, however, because it stores the human genome in quick access memory and will quickly search the database for alignments of DNA or protein sequences. The BLAT server is also unique because it contains one of the most up-to-date and non-redundant copies of the human genome.

Vertebrate Genome Databases

Animal genomics is split at the farm and pharmacy crossroads. Those traveling the farm road include the agricultural genome projects such as pig, cow, and chicken while the pharmacy caravan consists of mice and rats. The databases discussed in this chapter are listed with URLs in Table 9.2.

General Vertebrate Databases

Online Mendelian Inheritance in Animals (OMIA)

The Online Mendelian Inheritance in Animals is much like its human counterpart at NCBI because it organizes genes and disease information into one database. The difference lies in the way that OMIA catalogues genes for a large number of organisms (instead of just humans) and the fact that it is hosted at the ANGIS (Australian National Genomic Information Service) web server. A complete listing of organisms and genetic disorders/traits represented in the database can be found at the site and searched via the links on each of the subjects or through the search pages of the site. The records contain a description of the gene and the collection of references used to extract the gene information. It is an interesting way to browse genetic disorders across animal species.

ARKdb

The genetic mapping data for farmed animals is kept in the ARKdb at the Roslin Institute of Edinburgh. This database collects experimental information for ten species of farmed animals including cats (we have yet to see a cat farm) and fish. The database was originally developed to support the mapping projects for animals and can be accessed via the Anubis web program. This program provides a graphical interface to the genomes in the form of integrated graphical maps of cytogenetic traits, linkage group, and radiation hybrid data. The database is divided into four data types: references, loci, experiments' and maps. The references for each of the experiments are kept by the curators and linked to the Medline database at NCBI. The second data type, the genetic loci for the genomes, is sparse because of the limited amount of mapping data to date. The experimental methods and results are kept in the database, which can be queried to investigate the methods used to confirm the validity of the data. The fourth map data objects can be explored with the Anubis map viewer. The map viewer launches a interactive Java interface with built in controls or a dynamically drawn gif image maps. This mapping tool creates an excellent a graphical environment for browsing the various animal genomes.

Mouse

Mouse Genome Informatics

Mouse genome informatics is a research project of the Jackson Lab. In addition to being the primary source of mice research strains, the lab has developed a number of databases and other online resources. The databases include information on mouse genes, mapping data and homologies with other mammals. One of the more recent developments is the mouse phenomics database, a database of phenotypes linked to the sequence and stock databases. The Phenomics (phenotype + genomics) database contains records of phenotypic experiments led by the laboratory including a description of the

Table 9.2. Vertebrate Genome Databases

# Site Name	URL
1. ARKdb	http://www.thearkdb.org/
2. MGI – Mouse Genome Informatics	http://www.informatics.jax.org/mgihome/
3. MGD – Mouse Genome Database	http://www.informatics.jax.org/
4. GXD- Gene Expression Database	http://www.informatics.jax.org/
5. RGD – Rat Genome Database	http://rgd.mcw.edu/
6. Ratmap	http://ratmap.gen.gu.se/

experimental procedures and the results. Example experiments include the average weight of the strain over age, locomotive activity in the dark and taste preferences. The phenomic data is also linked to the records for each of the mice in the Jackson laboratories order catalogue. The catalogue information contains the genetic background of the strain, the dietary procedures and recommended research applications.

Mouse Genome Database (MGD)

Since 1994 the mouse genome database at the Jaskson Labs has provided online researchers with data for genetic, physical and experimental mouse genome maps. The MGD is one of the founding members of the Genome Ontology (GO) consortium and has been an active supporter of genomic data integration. The most recent releases of the MGD and its associated databases MGS (Mouse Gene Sequencing) and GXD (Gene Expression Database) have utilized the controlled vocabularies developed with the GO consortium to integrate with primary databases such as SWISS-PROT and NCBI.

The data in MGD includes genetic marker information, molecular segments, phenotypes and mapping data. The source data is updated daily from the various mouse sequencing and mapping projects from around the world. The database also links to relevant outside databases for homology comparisons such as BovMap, SheepBase, and PigBase. The homology information is taken directly from published literature and interpreted for use in three homology applications, the Oxford Grid, comparative map and Homology Map database views. Each of the applications creates different views of genomic comparisons, which are linked to the MGD primary data.

Gene Expression Database (GXD)

The Gene Expression Database at the Jackson Labs summarizes various types of experimental expression data from RNA *in situ* hybridization, northern blots, western blots and array data published for the mouse. The gene expression information has been integrated into the MGD with hyperlinks to the related information in each of the database records. The GXD query form is a primary access point into the GXD where users can formulate queries based on search parameters involving relevant criteria including tissue type, genetic locus or chromosome. This is facilitated by the collaborative efforts of the Gene Ontology project that has defined a controlled vocabulary for assigning the correct descriptions to genes. A Mouse Anatomical Dictionary Browser has also been created to guide researchers through the mouse anatomy and to help search for expression patterns from an anatomical vantage point. The GXD project has also developed the cDNA clone query form modeled after the GXD query form and Gene Expression Notebook for electronic data submission.

Rat

Rat Genome Database (RGD) and Ratmap

Much like the MGD, the Rat Genome Database (RGD) is built around the mapping and sequencing projects for the model organism. The Rat Genome project is expecting to have 3 to 4 times sequencing coverage by 2003 with full scale sequencing efforts starting in 2001. The RGD is a new rat resource that utilizes the mapping information from the Ratmap database and builds upon it with additional comparative mouse and human map data. The comparative analyses can be browsed at the Virtual Comparative Mapping page of the database. This mapping page is a starting point for browsing the chromosomes of each of the three species by clicking on the chromosome link. Following the link will guide you to a page of chromosome alignments, which display the syntenic regions of Mouse, Rat, and Humans and link to common genes of interest in each of the model organisms. The mapping information for the genome comprises most of the data on the website with sequences to be added as the project progresses. The majority of the sequences for the genome are from the BAC (Bacterial Artificial Chromosome), EST (Expressed Sequence Tag) and SSLP (Single Sequence Length Polymorphism) sequences used for the mapping project. The targeted links on each of the search pages have been posted in the left navigation bar to guide the researcher to related information and can be very useful when conducting research on the RGD site.

Plant Genome Databases

The plant databases are unique because there are two different methods to store descriptions of these rather large genomes. The first, used primarily for agricultural plants, is based on increasing the resolution of genetic maps, while the second is based on cataloguing data from sequencing efforts. The mapping-based genome databases are collected at agricultural research centers such as UK CropNet in the United Kingdom and the ARS Genome Database Resource at the Cornell University. These database centers have made use of the ACEDB genome database because of its flexibility and built-in ability to graphically display relationships between genetic-, physical-, and sequence-based information. See Chapter 10 for more information on AceDB. TIGR (The Institute for Genomic Research) is a source of much of the plant genome and transcriptome sequence information. The TIGR plant genome databases provide alternative sequencing-based databases that have been designed to receive the raw sequence and make it available for gene annotation and submission to the primary sequence centers. This section will profile some of these resources and how they may be used for online plant genetic research. See Table 9.3 for the URLs of the plant resources.

UK CropNet / ARS GDR

The United States and United Kingdom sister sites, ARS Genome Database Resources and UK CropNet , respectively, have one of the most extensive collections of plant genomes, the majority of which are running on the ACEDB genome database. The interface to the database is made available via the ACEDB web interface modules. The centers provide two interfaces to the ACEDB, the WebACE and ACE Browser, with two more (GFACE and WebACE2K) in development. While both databases have two viewing options, the WebACE browser is the default for the ARS GDR database. See Chapter 10 for more information on ACEDB. The databases also have a list of the all of the genomes made available via mirror (duplicate web page) on the server and links to the homepages where applicable. Some of the databases do not use the ACEDB as a database; the MaizeDB, for example, has adopted a Java-based database system from GeneLogic that launches a Java Applet in a new window to explore the database.

TIGR Plant Genomes

TIGR provides an alternative to the mapping project databases with its sequencing project databases. TIGR has created databases for rice, potato, tomato, and soybean to name a few. The difference in these databases is the focus or purpose of the information. Since TIGR is a sequencing facility, their focus is processing partial sequences for most of the plant genome projects from ESTs into full length genes. Along with the typical gene search via BLAST or keyword, the TIGR databases offer information about how the gene was assigned a function, whether through protein sequence alignment, homology to a known gene or other computationally derived methods. TIGR can be used to BLAST a particular genome's sequence without having to perform data presorts that would be required in larger databases. Depending on the project, TIGR may add additional value to the databases: The Potato Genome Project is the example of this, where part of the project is aimed at providing DNA array chips made in conjunction with the ESTs sequencing efforts.

Arabidopsis

Arabidopsis, the most popular plant genomics model organism was sequenced in 2000. With the sequencing project came a number of databases that offer different perspectives into the plants genome.

AGI - TAIR

The Arabidopsis Genome Initiative (AGI), the international collaboration responsible for the sequencing of first plant *Arabidopsis thaliana*, has created a comprehensive database and web interface to the genomic information of the model plant. The Arabidopsis Information Resource (TAIR) formerly known as AtDB is a collection of sequence, mapping and seed stock information for the Arabidopsis research community. TAIR has been designed to be an easy-to-use genome browser with all of the mapping, sequence, and community member information available on the model plant within three HTML pages (or clicks) of each other. At the time of this writing, some of the links to the research community are broken but for the most part all of the information is available.

TAIR uses an object oriented database system with Java classes to create the on-the-fly graphical mapping capabilities that can be seen when browsing the chromosomes on the site. The Map Viewer is one of the most advanced web-based graphical tools that you will run across on the web. It allows simultaneous viewing and searching for

Table 9.3. Plant Genome Databases

#	Site Name	URL
1.	UK CropNet	http://ukcrop.net
2.	ARS - Genome Database Resource	http://ars-genome.cornell.edu/
3.	TIGR Plant Genomes	http://www.tigr.org/tdb/
4.	AGI – Arabidopsis Genome Initaitive	http://www.arabidopsis.org/agi.html
5.	TAIR – The Arabidopsis Information Resource	http://www.arabidopsis.org
6.	MATDB	http://www.mips.biochem.mpg.de/proj/thal/
7.	PENDANT	http://pedant.mips.biochem.mpg.de/
8.	AGR	http://ukcrop.net/agr/
9	TIGR AtDB	http://www.tigr.org/tdb/e2k1/ath1/
10.	Banana Genomic News	http://www.genomeweb.com/articles/view-article.asp?Article=20017191138

clones, markers or genes on a variety of genomic maps including Recombinant Inbred (RI) maps, Classical Genetics Maps and the AGI sequence based map. At any point along the search you may investigate a feature by clicking it. This will either zoom into or call a summary page, which displays associates to other data types, the history of the feature and comments.

MATDB

MIPS *Arabidopsis thaliana* Data Base (MATDB) has two browser interfaces to the annotated sequence information for Arabidopsis. Both lead to a list of the clones used to sequence the genome but they start at different points. The graphical viewer starts at the chromosome and can be explored by clicking on the chromosome of interest. This will lead you to a list of clones and eventually to a record for an individual clone. This can also be done with the list view where all of the clones are listed by name. The interesting analysis begins when the gene is chosen from the list of genes on the clone. The first page displays information gleaned from the MATDB with links to the alternatively annotated TIGR AtDB and PENDANT (Protein Extraction, Description, and Analysis Tool). PENDANT is the mechanism by which the genes are annotated for the database; it can perform a number of different types of analysis on the sequence such as protein translations, PFAM analysis, BLOCKS, 3D structure and BLAST to various other model organisms. Overall MATDB provides an informative resource for genetic and protein work on *Arabidopsis*.

AGR Arabidopsis Genome Resource (UK CropNet)

The AGR makes extensive use of the ACEDB system of data archiving. The interface to the genome uses the webACE browser with some modifications that add functionality to the database. The additional features include the capability to view the sequence in a graphical format, printouts of the translated sequence, exon descrptions, and BLAST search capabilities. The server has also included a BLAST server to search for sequence similarities for all of the other ACEDB's on the server. Note that the BLAST search is not integrated into the BLAST function—you must cut and paste the sequence into the text box on the BLAST page. An important aspect about the ACEDB databases is their tight linkage to genetic mapping data, especially with plant genomes. In addition, the database contains information on the availability of germplasm and images linked to phenotypes. In the future more linkage between genetic and genomic data will be added to the database, including microarray and additional expression profiling.

TIGR's AtDB

The TIGR *Arabidopdid thaliana* Genome Annotation Database is the collection of all the genes in the *Arabidopsis* genome. With a variety of search options such as Gene Name Search, Clone Search, and Locus Search, the database provides a very complete record search of the genes in *Arabidopsis*. This database is also equipped with a chromosome browser similar to the chromosome browsers on the MATDB and TAIR. With the linked features you will find the browsing experience familiar and intuitive but the end results different. The summary page for a TIGR search is a table of attributes associated with the potential gene of interest with sequence alignment images at the bottom of the page. The sequence alignments are used to determine the putative function of the gene.

Banana Genomics

At the time of this writing the banana genome project has been proposed as one of the next plants to be sequenced. With a genome of five hundred thousand base pairs, it is one of the smallest plant genomes in existence and could be sequenced within the next five years.

Insect Genome Databases

The insect genome databases (see Table 9.4) begin with the century old model organism *Drosophilia melanogaster*. The well-studied fruit fly has a legacy of genetic research from its origin as a tool for genetic recombination research to the more recent studies of the developmental stages of the embryo. What is unique about *Drosophilia* research is that much of it is documented and the years of documentation have been added to the nucleotide sequencing databases to create some of the most comprehensive databases to date.

Drosophilia

Flybase

The Flybase database represents the entire Drosophilidae family but focuses on *Drosophila melanogaster*. Full mirrors of the website are available at Berkeley, Harvard and Bloomington, Indiana. Partial mirrors are available throughout Europe, the UK, France, Taiwan, and Japan. Flybase is very advanced in its user interface capabilities with different start page menus specific to the type of research to be browsed. There is also a preferences panel that can be linked to a personalized account to keep a browsing history across different sessions. In the preferences table you can choose your preferred searching, browsing and reports options after a personal account has been established.

Gene Reports come in three varieties. The synopsis report displays the summaries of each of the major features of a database entry. Additional information is available via hyperlinks to other sections related to the entry. The abridged version displays brief descriptions of the record information, with just enough information to know if you have found the correct entry or not, while the full text report displays all of the available information for the particular record.

Considerable time and effort has been spent adding cross-referenced data to the gene records. This can be seen in the wide variety of additional links to the Protein and Transcript pages that display the protein product of a gene, non-drosophila homologues and protein domains originating from the nucleotide sequence. These pages also display the expression patterns of the gene using the controlled vocabularies of other genomes projects to facilitate a consistent method of searching the entries. Genome annotation or assigning nucleotide sequence to proteins and proteins to function has been integrated via the GADFly project. GADFly is part of the Flybase network but is not on the same server as the Flybase and serves as the genome annotation tool for Flybase. The database is hosted at the Lawrence Berkeley National Labs and is connected to all of the available databases.

Table 9.4. Insect Genome Databases

# Site Name	URL
1. FlyBase	http://flybase.bio.indiana.edu/
2. The Interactive Fly	http://sdb.bio.purdue.edu/fly/aimain/1aahome.htm
3. Flybrain	http://flybrain.neurobio.arizona.edu/Flybrain/html/index.html
4. FlyView	http://flyview.uni-muenster.de/
5. FlyMove	http://flyview.uni-muenster.de:8080/
6. Mosquito Genome Page	http://klab.agsci.colostate.edu/

The Interactive Fly

The Interactive Fly is a website of all things related to the genes involved in the development of Drosophila. The database contains a complete alphabetical listing of the genes involved in the entire process, a cross reference via gene name with FlyBase, lists of genes according to biological function and genes involved in organ and tissue development. There is also an image archive and a search engine.

The records for the gene entries include a comprehensive biological overview of the gene products role in the fly development process. The database record also contains an interpretation of the gene functions on the molecular level, descriptions of the gene transcription and protein interactions. The page contains an impressive amount of written information on the Drosophila developmental process.

Flybrain

The atlas of the *Drosophila* brain contains a huge image library of images, a 3D project with VRML and QuickTime interactive displays of the brain at different stages of development. The database also includes a genetic dissection of the brain with a mutation analysis database and enhancer trap lines, which show gene expression in the brain for a set of genes. Most of the images are organized into atlases and are linked based on the area of the brain being researched at the time. There also links to developmental studies publications about the *Drosophilia* brain or links to the images for the publications. In keeping with the Flybase project the Flybrain project has developed a controlled vocabulary but it is unclear if it is consistent with the other controlled vocabulary projects.

FlyView

Flyview is an image database of 3,700 *Drosophila* images in all stages of growth with a text field for searches accompanying each record. There is also a call for all enhancer trap images to the *Drosophila* community, they would like images of published and unpublished enhancer trap lines to be available for public display.

FlyMove

Yet another development and image database for *Drosophila*, Flymove, is a basic overview of the fly with quicktime images and short explanatory text. The flatfile database has two entry points, a graphical entry point where the embryo can be inspected visually and the organs clicked for more information or a text mode where all of the information is displayed in HTML tables.

Mosquito

Mosquito Genomics

The Mosquito genome page contains databases of various Mosquito genomes in the ACEDB format. From the homepage online researchers are able to BLAST or keyword search sequence data from the yellow fever mosquito, the Asian tiger mosquito, the Eastern tree hole mosquito, the Northern house mosquito and the malaria mosquito. This database is the precursor to a genome project proposal for the yellow fever mosquito (*Aedes aegyti*) for which there is additional information on the web page.

Fungus Genome Databases

As the first eukaryotic organism to be completely sequenced yeast is the foundation for fungal genetics. The yeast genome community finished sequencing the genome in 1996 and has since added information about the yeasts proteome and transcriptome to create a very complete picture of this model organism. The online resources available for yeast begin at the Saccharomyces Genome Database (SGD) where genomic information from protein structures to gene expression have been collected and published since the inception of the sequencing project. Other available resources are the yeast proteome database, an excellent source of curated yeast protein data, and the MIPS yeast genome database, a genome database with extensive tables of protein sequences derived from the primary nucleotide data. (See Table 9.5 for URLs)

Yeast

Saccharomyces Genome Database (SGD)

The Saccharomyces Genome Database (SGD) collects and organizes the genomic information for the yeast *Saccharomyces cerevisiae*. The SGD has been a leader in database technologies and is one of the first databases to adopt the controlled vocabularies of the Gene Ontology consortium. The SGD contains multiple entry points into the yeast genome including hyper linked genetic and physical maps, BLAST, FASTA, and homology search pages. The new features of the online database are the Function Junction a portal to annotated gene function, and Expression Connection, a web entry point into the yeast gene expression data.

Table 9.5. Fungus Genome Databases

#	Site Name	URL
1.	SGD - Saccharomyces Genome Database	http://genome-www.stanford.edu/Saccharomyces/
2.	YPD – Yeast Proteome Database	http://www.proteome.com/
3.	MYPD – MIPS Yeast Proteome Database	http://www.mips.biochem.mpg.de/proj/yeast/

The Function Junction is a distributed search system that allows researchers to search multiple online functional analysis databases from a single entry point. The results of which can be used to help designate gene functions or to compare functions of genes from other organisms to those in yeast. The Expression Connection queries multiple databases for microarray expression data and displays the expression pattern for the gene of interest. The user may choose various graphical displays of the information including plots for serial experiments. Together, these tools add significant value to the vast amount of sequence data from the nucleotide sequencing project.

Yeast Proteome Database (YPD)

The Yeast Proteome Database, together with PombeDB (*Schizosaccharomyces pombe*, fission yeast) and WormDb (*Caenorhabditis elegans*), make up the BioKnowledge Library at Proteome. This commercial website employs researchers to add additional information to the yeast proteome from published scientific literature. The result of this additional curation is a Protein Report with a summary of the function, localization, and interaction of the protein within the organism. To access the data the website has searches based on keyword, gene name, and protein name that are also connected to a synonym checking database to help focus query terms. The BioKnowledge web site requires login but is available free to academic researchers.

MIPS Yeast Genome Database (MYGD)

The Munich Information center for Protein Sequences compiled its own version of the Yeast genome using the DNA and protein sequences from the primary sequence databases. The MYGD contains a simple and easy-to-use interface to the Yeast genome with search functions for the entire genome complete with sub-searches, maps of the chromosomes and individual analysis and interaction tables, which contain data about yeast protein interactions.

Invertebrate Genome Databases

The hermaphroditic nematode *Caenorhabis elegans* is a favorite model organism for genetics in part because the discrete number of cells that can be traced throughout its life cycle. The *C elegans* sequencing project at the Sanger Centre has developed a number of database tools to accompany the sequencing project including an AceDB database WormBase and the WormPep protein database. Additional information about *C. elegans* can be found at the Proteomes BioKnowledge website. All website URLs for the databases of this chapter can be found in Table 9.6.

C. elegans

WormBase

Is based on the AceDB and is an excellent example of the flexibility of the AceDB database. WormBase differs from the AceDB project of providing mapping and sequencing information because it incorporates information from literature to give a holistic view of the worm. The genome website contains multiple search capabilities ranging from a simple sequence search to an advanced search using the native ACE query language. Also built into the web application are search tools for nucleotide and protein sequences, a cell lineage

Table 9.6. Invertebrate Genome Databases

#	Site Name	URL
1.	WormBase	http://www.wormbase.org
2.	AceDB	http://www.acedb.org
3.	Sanger *C. elegans* Project	http://www.sanger.ac.uk/Projects/C_elegans/
4.	WormPep	http://www.wormbase.org
5.	Dictybase	http://www.dictybase.org

browser and an expression search page. The project is also developing a GO (Gene Ontology) *C. elegans* browser with the tools provided from the GO project. (See chapter 9 for more information about the GO project). One helpful feature built into some of the search pages is a query builder. The expression database requires that three parameters submitted into the database via three lists. After the user has chosen the parameters and submitted the search to the page a new page will load with the search results (should there be any) and a textbox with the ACE query that was just performed. From this page the search parameters can be altered to achieve the desired results. This is also an excellent way to learn the ACE query language syntax without having to write the code.

AceDB

A C. *Elegans* Data Base (AceDB) is a database designed flexible enough to be customized for any genome project yet robust enough to hold a genome's worth of sequence data. See chapter 10 for more information on the AceDB database. Originally designed for the *C. elegans* genome project the database has been adapted for numerous plant genomes, animal genomes and other non-biological database projects. The *C. elegans* sequencing project still uses the database but not through the AceDB project website as one might expect. Instead the information for the *C. elegans* genome can be found at

WormBase and information about AceDB software can be found at the ACEDB.org web site. This website also contains information on how to set up and administer the free database software and is the place to go for expert AceDB help. (See Chapter 10 for more information)

C. elegans AceDB Browser at the Sanger Centre

The Sanger Centre is responsible for the sequencing of the *C. elegans* genome with the real AceDB used for C Elegans. The project houses a database with a simple web front end to the ACEDB browser at WormBase, meaning that any data submitted to the Sanger Center project page is transferred and run on the WormBase page. The full version of the genome browser resides at the WormBase site (full version meaning text and map images).

WormPep

WormPep a database of *C. elegans* protein sequences is also available at the Sanger center. It accepts either an accession number or keywords and returns the protein sequence in FATSA format. WormPep has been integrated into the WormBase database.

Dictyostelium

Dictybase

The peculiar soil amoeba *Dictyostelium discoideum* has been targeted for genomic sequencing and should be sequenced within a few years of this writing. The genome project information at www.dictybase.org contains information about the project and links to available resources.

Bacterial Genome Databases

Bacterial genome databases have been fueled by both the need to understand more about the diseases caused by pathogenic strains of bacteria and to better utilize bacteria as the molecular toolboxes of the laboratory. Online bacterial strain databases, genome project databases and comparative genome maps make up a few of the resources available for researchers on the Internet. The Institute for Genome Research, a leader in prokaryotic genome sequencing efforts, has complied a database of all known bacterial genomes and made it available via a web interface at the Comprehensive Microbial Resource page of their web site. Additionally, numerous centers around the world have copies of the *Escherichia coli* genome and have independently developed secondary databases each with a unique analysis of the bacterial genome. Below is a listing of general bacterial genome resources along with a few descriptions of the *E. coli* resources available on the Internet (for URLs, see Table 9.7).

General Bacterial Genome Databases

World Data Centre for Microorganisms (WDCM)

The WFCC-MIRCEN World Data Centre for Microorganisms is a portal to the collections of microbial cultures. The website contains links to stock centers, microbial collections and cell banks around the world. Keyword search pages have also been made available to assist researchers. After choosing an area to search, the database prompts the user for a key word and returns a list of the possible matches. Clicking on a hyper linked record name opens a new page with a description of the item and information about the availability of the culture and contact information for retrieving material from the collection.

Comprehensive Microbial Resource (CMR)

The Institute for Genome Research (TIGR) has developed a database of all the bacterial genomes sequenced to date and has made it available online at the Comprehensive Microbial Resource (CMR) page of the TIGR website. The core of the website, the Omninome database, contains the sequence and annotation of the DNA and protein sequences for the microbial genomes, information about the microbes themselves and genome wide information about the composition of the genome as a whole (GC content etc.). A key feature of the CMR is its ability to search for molecular properties such as GC content, Ip (Iso-electric point), molecular weight, common name, gene symbol, and role across the genomes. The CMR is a unique cross-genome database querying tool for online researchers. To add depth to the search results each of the protein sequences in the Omniome has been searched against the database and the results kept as paths to related records. The precompiled paths to other proteins or genes in related genomes are made available on the search results pages of the database as links to related database entries. This appears as smarter search results because of the built in link-out function.

In addition to pre-compiling the database, TIGR has create a number of large scale sequence analysis tools for the genomes such as the Multi-Genome Query Page, a query building application, and the Restriction Digest Tool which displays both the cut sites for the genome and a digitally created gel image of the band sizes after digest. The CMR also contains pages of multi-genome lists by categories, which file the genes, RNA or the proteins of an organism so that they may be browsed and downloaded. The third type of tool the Multi-Genome Analysis tool extracts information from the genomes, which could be useful for researchers wanting to compare the overall characteristics of the genome. An example of the tool is the Genome v/s Genome Protein Hits tool that displays the number of proteins in each organism with a corresponding protein in all the other organisms. In addition to the built-in tools and applications, the CMR provides Genome Pages for each of the microbes in the Omninome database.

The Genome Pages contain information about the project behind the sequencing of the organism links to search pages, the lists of the genes in the organism and related Internet resources

Microbial Genome Database (MGDB)

The goal of the MGDB is to collect and analyze all of the genes and putative genes for the sequenced microbial organisms into homologous – orthologous clusters. The genes are compared and the similarities stored in the homologous gene cluster table to be later used for the gene analysis functions. From this table researchers are able to investigate the genomes with the searching and analysis functions built into the MBGD database. The principal function of the database is the Homologous Gene Table Creation function. This function clusters the complete gene set of the genomes chosen at the search pages of the site and returns a table of the organisms to which the clusters belong and a synopsis of the genes. The clusters can be clicked to view the details of the analysis and to further investigate the groups of genes.

HOBACGEN: Homologous Bacterial Genes Database

HOBACGEN is a database of the bacterial protein sequences organized into families. The database contains all nucleotide and protein sequences from all species of bacteria and yeast in the SWISS-PROT + TrEMBL databases. Access to the database is available via a downloadable Java GUI or via a web interface. In either case a query is submitted to the HOBACGEN database and a list of the protein families of interest are returned. The resultant list of families can be used to create multiple sequence alignments and phylogenetic trees. The sequence alignments are done with the CLUSTALW software and the alignments with the BIONJ software.

Table 9.7. Bacterial Genome Databases

#	Site Name	URL
1.	WDCM – World Data Centre for Microorganisms	http://wdcm.nig.ac.jp/
2.	CMR – Comprehensive Microbial Resource (TIGR)	http://www.tigr.org/tigr-scripts/CMR2/CMRHomePage.spl
3.	MGDB – Microbial Genome Database	http://mbgd.genome.ad.jp/
4.	HOBACGEN – Homologous Bacterial Genes Database	http://www.hgmp.mrc.ac.uk/Registered/Option/hobacgen.html
5.	EcoGene	http://bmb.med.miami.edu/EcoGene/EcoWeb/
6.	EcoWeb	http://bmb.med.miami.edu/EcoGene/EcoWeb/
7.	ECGC - *E. Coli* Genome Center	http://www.genome.wisc.edu/
8.	Colibri	http://genolist.pasteur.fr/Colibri/

E. coli

EcoGene Web Site: EcoWeb

EcoGene is the source of re-annotated *E. coli* protein sequence for SWISS-PROT and Colibri databases. Genbank provides the source protein sequences for the EcoGene database, which are then reanalyzed for the correct start sites, the correct frame shifts, reconstructed protein sequences and gene predictions. The resultant database is an alternative representation of the *E. coli* genome. This database is continually revised as technologies progress and literature is published to contain the most accurate gene annotation information. The data is available at the Colibri database (see table below) where it is integrated to other genome information and at the EcoWeb web site. This web site also contains tables of the latest database releases available for browsing. These tables have been annotated and include links to primary sequence sources.

E. Coli Genome Center (ECGC)

The home for the *E. Coli* genome project is at the University of Wisconsin, Madison. This web page contains multiple tools for browsing the genomes of different *E. coli* strains and is the location of various pathogenic *E. coli* genome sequencing projects. The tools section of the database launches Java applet based applications for interactive genome browsing. These can be used to view a recently annotated version of the genome by clicking on a region of interest in a linear or circular depiction of the *E. coli* chromosome. The page also contains functional genome analysis information available for download and houses an *E. coli* BLAST server.

Colibri

Colibri contains all of the protein and nucleotide sequence data from the *Escherichia coli* strain K-12 linked to functional annotations. The database has an Internet component available on the web for searching the sequences either by keyword or using the graphical representation of the chromosome map. Clicking on a region of the chromosome will open a frame of the web page with a picture of the genes, as they are located on the strand of DNA. Clicking on the gene of interest will load another browser frame with the record information about the gene including its accession number, description and cross-references to the source databases SWISS-PROT and GenBANK. The Colibri database is built upon a database system originally used for the genome sequence project for *Bacillus Subtilis* called the SubiList.

Table 9.8. Organelle Databases

#	Site Name	URL
1.	AmmtDB	http://bio-www.ba.cnr.it:8000/BioWWW/#AMMTDB
2.	GoBASE	http://megasun.bch.umontreal.ca/gobase
3.	HvrBase	http://bio-www.ba.cnr.it:8000/srs6
4.	MitAln	http://bio-www.ba.cnr.it:8000/srs6
5.	MitBase	http://www3.ebi.ac.uk/Reaseach/Mitbase/mitbase.pl
6.	MitBase Pilot	http://www3.ebi.ac.uk/Reaseach/Mitbase/mitbase.pl
7.	MitoDat	http://www-lecb.ncifcrf.gov/mitoDat
8.	MITOMAP	http://www.gen.emory.edu/mitomap.html
9	MitoNuc	http://bio-www.ba.cnr.it:8000/srs6
10.	MITOP	http://websvr.mips.biochem.mpg.de/proj/medgen/mitop
11.	PLMItRNA	http://bio-www.ba.cnr.it:8000/srs6

Organelle Databases

Organelle sequences are abundantly available in the primary sequence repositories. This has made them a favorite for phylogenetic studies both across and within species. Much of the additional information present in specialized databases complements studies in comparative genomics. For a listing of URLs for the following sites, see Table 9.8.

Organelles

Phylogenetics has been the driving force for much of the organelle database development on the web. Included with the requisite annotation by expert research labs, often includes information such as the geographic origin, collection method or other distinguishing characteristics of an organism's mitochondria. The mitochondrial databases are divided among four large categories that include humans, plants, vertebrates and fungus. The outlier for the group is the set of human databases because of the additional information about human and great ape phylogenetic data and because of the interest in mitochondrial genetic mutations. In addition to the information about the mitochondrial genome, the databases often include DNA sequences for mitochondrial genes that have been incorporated into the host genome. These are often incorporated into the same dataset where available.

Mitochondria

Mitochondrial databases resemble the genome browsers of the larger databases with their ability to search and compare genomic sequences across species. The difference lies in the size and breadth of the database and the quality of the individual entries. MitoDat, MITOMAP, MitOP and AMmtDB concentrate on human and animal mitochondrial disease mutations. These databases query the central repositories and reformat the data in either aligned or annotated flatfile formats. (See the table of the URLs and query interfaces below.) MitoBase, an international consortium of mitochondria specialists, is responsible for the curation of various kingdoms of mitochondrial data including humans, fungi, vertebrates and plants. One of the primary concerns of the consortium is to develop a common language to describe the data in the database repositories and to bring together additional information via links to primary literature and the cross referencing of related data object. The consortium has also developed MitoBase Pilot, a database of nuclear genes of the mitochondria. This database again focuses on the accurate description of the mitochondrial genes and adds additional content specific links to the existing gene entries. One of the most powerful aspects of relational databases such as these is the ability to develop complex search queries. For this reason the databases have been developed around the SRS query system and all can be accessed via SRS servers at various locations. SRS, Sequence Retrieval System, is discussed in Chapter 10.

Mitochondria and Beyond

GOBASE, an organelle database that has been slated to cover mitochondria, chloroplast and model eubacteria sequences, is unique amongst the organelle databases. While the others focus on either human mitochondria or variants of mtDNA sequence within a species, GOBASE provides a genomic view of mitochondria. With expert annotation and monthly queries of the primary sequence repositories, GOBASE boasts the most up to date and accurate mitochondrial data available on the web. By 2001 the database curators hope to include chloroplast, cyanobacteria and -proteobacteria in the database. Another feature of this database is its unique querying system, which requires additional learning but uses essentially the same logical steps for retrieving database information on the web.

Virus Genome Databases

In previous sections of the chapter we have discussed controlled vocabularies and their role in the data archiving process. Virology has its own version of a controlled vocabulary for the taxonomic naming of the many different virus species. This vocabulary has been an important organizational tool for the virus databases around the world. Because of their size, virus genomes (together with organelles) make up the largest number of genomes in the primary sequence databases. This means that single virus genome resources are not as common, and webs sites are instead dedicated to either entire families of viruses or to all viruses as a whole. Virus genome pages also tend to lack the mapping data usually associated with genome projects, but this is made up for by the amount of nomenclature involved in the curation process. The following descriptions are some of the major virus data repositories available on the Internet. URLs for the websites can be found in Table 9.9.

General Virus Genome Databases

All the Virology on the WWW

This site is the online portal to all the virology information on the Internet. It has information and links to all the virology related websites and is built in conjunction with the Big Picture Book of Virology, a guide to virus taxonomy. This page is a great resource for the general public and the virologist who needs additional information on how to access viruses on the web.

International Committee on Taxonomy of Viruses (ICTV)

Nomenclature is a recurrent theme in database development and has been a topic amongst virologists since 1966 when the ICTV was established. The mission of the ICTV is to perpetuate the systematic naming of viruses based on virion characteristics instead of the disease phenotype witch they induce (e.g. mosaic, pox). The ICTYdb is a series of web pages, which help to place the viruses in families and genera. This online resource is used by virologists to place species into the correct families and is a definitive list of the correct names of the listed species. This list is used as the source of the controlled vocabulary of virus names for many of the databases, which cross-reference virus names with sequence data.

Plant Viruses Online

The Plant Viruses Online website from Australia is a result of the Virus Identification Data Exchange (VIDE) project from which the data is derived. The pages are long lists of viruses and viroids with cross references to their plant host species. The pages are connected by hyperlink from host to virus pathogen to facilitate online browsing. The pages do not have a search engine and must be browsed manually.

Table 9.9. Virus Genome Databases

#	Site Name	URL
1.	All the Virology on the WWW	http://www.virology.net/
2.	ICTV – International Committee on the Taxonomy of Viruses	http://www.ncbi.nlm.nih.gov/ICTV/
3.	Plant Viruses Online	http://image.fs.uidaho.edu/vide/refs.htm
4.	VIDA	http://www.biochem.ucl.ac.uk/bsm/virus_database/VIDA.html
5.	HIV Sequence Database	http://hiv-web.lanl.gov/

VIDA

VIDA is a database of animal virus genome open reading frames for the *Herpesviridae*, *Coronaviridae* and *Arteriviridae* families. The database contains genome sequence data from the primary sequence databases and uses a controlled taxonomic vocabulary (created by The International Committee on Taxonomy of Viruses) to organize the open reading frames so that they may be used to compare the genomic information across species. The database performs a number of sequence similarity searches to determine homologous shared protein sequences. The searches start with single domain matches and build to eventually compare all of the domains for a given set of sequences. The resultant data are homologous families of proteins named HPFs. These are stored in the database as HPF entries and are the primary data type of the VIDA database. Links to primary databases have been added to the individual sequence records along with information about the protein structure of the sequence along with functional annotation where possible.

HIV

HIV Sequence Database

Unlike the genome databases of other organisms, virus genome pages typically do not have mapping data but in some cases do have extensive sequence comparison. The HIV Sequence database contains pages with precompiled sequence alignments of all HIV-1, HIV-2/SIV and SIV-agm genes. Downloadable analysis tools and two sequence search interfaces are also available. The sequence search interfaces are easy to use and give extensive gene alignment information. There is also a search page from which one can identify mutations and possible variants of the virus. The HIV sequence database is one example of a virus web page and database dedicated to a smaller subset of virus genomes.

Further Reading

Human Databases

Cotton, R.G., McKusick, V., and Scriver, C.R. 1998 The HUGO Mutation Database Initiative. Science. 279(5347):10-1.

Frezal J. 1998. Genatlas database, genes and development defects. Review. C R Acad Sci III. 10:805-17.

Hamosh. A,, Scott. A.F., Amberger ,J., Valle, D., and McKusick, V.A. 2000. Online Mendelian Inheritance in Man (OMIM). Hum Mutat. 1: 57-61.

Letovsky, S.I., Cottingham, R.W., Porter, C.J., and Li, P.W. 1998. Related Articles GDB: the Human Genome Database. Nucleic Acids Res. 1: 94-99.

Rebhan, M., Chalifa-Caspi, V., Prilusky, J., and Lancet, D. 1998. GeneCards: a novel functional genomics compendium with automated data mining and query reformulation support.

Bioinformatics. 8: 656-664.

Stoesser, G., Baker, W., Van den Broek, A., Camon, E., Garcia-Pastor, M., Kanz, C., Kulikova, T., Lombard, V., Lopez, R., Parkinson, H., Redaschi, N., Sterk, P., Stoehr, P., and Tuli, M.A. 2001. The EMBL nucleotide sequence database. Nucleic Acids Res. 29: 17-21.

Talbot, C.C. Jr., and Cuticchia, A.J. 1999. Human Mapping Databases, Current Protocols in Human Genetics. John Wiley & Sons, Inc, New York. p. 1.13.1-1.13.12,

Wheeler, D.L., Church, D.M., Lash, A.E., Leipe, D.D., Madden, T.L., Pontius, J.U., Schuler, G.D., Schriml, L.M., Tatusova, T.A., Wagner, L., and Rapp, B.A. 2001. Database resources of the National Center for Biotechnology Information. Nucleic Acids Res. 1:11-16.

The Computational Biology and Informatics Laboratory. DoTS: a database of transcribed sequences for human and mouse genes. 2001. Center for Bioinformatics, University of Pennsylvania. http://www.cbil.upenn.edu/downloads/DoTS/

Vertebrate Databases

Blake, J.A,, Eppig, J.T., Richardson, J.E., Bult, C.J., and Kadin, J.A. 2001. The Mouse Genome Database (MGD): integration nexus for the laboratory mouse. Nucl. Acids. Res. 29: 91-94.

Hu, J., Mungall, C., Law, A., Papworth, R., Nelson, J.P., Brown, A., Simpson, I., Leckie, S., Burt, D.W., Hillyard, A.L., and Archibald, A.L. 2001. The ARKdb: genome databases for farmed and other animals. Nucleic Acids Res. 1:106-110.

Ringwald, M., Eppig, J.T., Begley, D.A., Corradi, J.P., McCright, I.J., Hayamizu, T.F., Hill, D.P., Kadin, J.A., and Richardson, J.E. 2001. The Mouse Gene Expression Database (GXD). Nucl. Acids. Res. 29: 98-101

Plants

Dicks, J., Anderson, M., Cardle, L., Cartinhour, S., Couchman, M.,

Davenport, G., Dickson, J., Gale, M., Marshall, D., May, S., McWilliam, H., O'Malia, A., Ougham, H., Trick, M., Walsh, S., and Waugh, R. 2000. UK CropNet: a collection of databases and bioinformatics resources for crop plant genomics. Nucleic Acids Res. 1:104-107.

Huala, E., Dickerman, A.W., Garcia-Hernandez, M., Weems, D., Reiser, L., LaFond, F., Hanley, D., Kiphart, D., Zhuang, M., Huang, W., Mueller, L.A., Bhattacharyya, D., Bhaya, D., Sobral, B.W., Beavis, W., Meinke, D.W., Town, C.D., Somerville, C., and Rhee, S.Y. 2001. The Arabidopsis Information Resource (TAIR): a comprehensive database and web-based information retrieval, analysis, and visualization system for a model plant. Nucleic Acids Res. 1: 102-105.

Mewes, H.W., Frishman, D., Gruber, C., Geier, B., Haase, D., Kaps, A., Lemcke, K., Mannhaupt, G., Pfeiffer, F., Schuller, C., Stocker, S., and Weil, B. 2000. MIPS: a database for genomes and protein sequences. Nucleic Acids Res. 1: 37-40.

Rhee, S.Y., Weng, S., Bongard-Pierce, D.K., Garcia-Hernandez, M., Malekian, A., Flanders, D.J., and Cherry, J.M. 1999. Unified display of Arabidopsis thaliana physical maps from AtDB, the A.thaliana database. Nucleic Acids Res. 1: 79-84.

Insect

Armstrong, J.D., Kaiser, K., Muller, A., Fischbach, K.F., Merchant, N., and Strausfeld, N.J. 1995. Flybrain, an on-line atlas and database of the Drosophila nervous system. Neuron. 1: 17-20.

Brody T. 1999. The Interactive Fly: gene networks, development and the Internet. Trends Genet. 8: 333-334.

Gelbart, W.M., Crosby, M., Matthews, B., Rindone, W.P., Chillemi, J., Twombly, S.R., Emmert, D., Ashburner, M., Drysdale, R.A., Whitfield, E., Millburn, G.H., de Grey, A., Kaufman, T., Matthews, K., Gilbert, D., Strelets, V., and Tolstoshev, C. 1999. The FlyBase database of the Drosophila Genome Projects and community literature. The FlyBase Consortium. Nucl. Acids. Res. 27: 85-88.

Janning W. 1997. FlyView, a Drosophila image database, and other Drosophila databases. Semin Cell Dev Biol. 1997. 5: 469-475.

Fungus

Ball, C.A., Jin, H., Sherlock, G., Weng, S., Matese, J.C., Andrada, R., Binkley, G., Dolinski, K., Dwight, S.S., Harris, M.A., Issel-Tarver, L., Schroeder, M., Botstein, D., and Cherry, J.M. 2001. Saccharomyces Genome Database provides tools to survey gene expression and functional analysis data. Nucleic Acids Res. 2001. 1: 80-81.

Costanzo, M.C., Crawford, M.E., Hirschman, J.E., Kranz, J.E., Olsen, P., Robertson, L.S., Skrzypek, M.S., Braun, B.R., Hopkins, K.L., Kondu, P., Lengieza, C., Lew-Smith, J.E., Tillberg, M., and Garrels, J.I. 2001. YPD, PombePD and WormPD: model organism volumes of the BioKnowledge library, an integrated resource for protein information. Nucleic Acids Res. 1: 75-79.

Mewes, H.W., Frishman, D., Gruber, C., Geier, B., Haase, D., Kaps, A., Lemcke, K., Mannhaupt, G., Pfeiffer, F., Schuller, C., Stocker, S., and Weil, B. 2000. MIPS: a database for genomes and protein sequences. Nucleic Acids Res. 2000 1: 37-40.

Invertebrate

Stein, L., Sternberg, P., Durbin, R., Thierry-Mieg, J., and Spieth, J. 2001. Related Articles WormBase: network access to the genome and biology of Caenorhabditis elegans. Nucleic Acids Res. 1: 82-86.

Walsh, S., Anderson, M., and Cartinhour, S.W. 1998. ACEDB: a database for genome information. Methods Biochem Anal. 39: 299-318.

Bacteria

Blattner, F.R., Plunkett, G. 3rd, Bloch, C.A., Perna, N.T., Burland, V., Riley, M., Collado-Vides, J., Glasner, J.D., Rode, C.K., Mayhew, G.F., Gregor, J., Davis, N.W., Kirkpatrick, H.A., Goeden,

M.A., Rose, D.J., Mau, B., and Shao, Y. 1997. The complete genome sequence of *Escherichia coli* K-12. Science. 277(5331): 1453-1474

Medigue, C., Viari, A., Henaut, A., and Danchin, A. 1993. Colibri: a functional data base for the *Escherichia coli* genome. Microbiol Rev 3: 623-654

Peterson, J.D., Umayam, L.A., Dickinson, T., Hickey, E.K., and White, O. 2001. The Comprehensive Microbial Resource. Nucl. Acids. Res. 29: 123-125.

Perrière, G., Duret, L., and Gouy, M. 2000. HOBACGEN: database system for comparative genomics in bacteria. Genome Res. 10: 379-385.

Rudd, K.E. 2000. EcoGene: a genome sequence database for *Escherichia coli* K-12. Nucleic Acids Res. 1: 60-64.

Organelle

Attimonelli, M., Altamura, N., Benne, R., Brennicke, A., Cooper, J.M., D'Elia, D., Montalvo, A., Pinto, B., De Robertis, M., Golik, P., Knoop, V., Lanave, C., Lazowska, J., Licciulli, F., Malladi, B.S., Memeo, F., Monnerot, M., Pasimeni, R., Pilbout, S., Schapira, A.H., Sloof, P., and Saccone, C. 2000. MitBASE: a comprehensive and integrated mitochondrial DNA database. The present status. Nucleic Acids Res. 1: 148-152.

Burckhardt, F., von Haeseler, A., and Meyer, S. 1999. HvrBase: compilation of mtDNA control region sequences from primates. Nucleic Acids Res. 1: 138-142.

Kogelnik, A.M., Lott, M.T., Brown, M.D., Navathe, S.B., and Wallace, D.C. 1998. MITOMAP: a human mitochondrial genome database—1998 update. Nucleic Acids Res. 1:112-115.

Lanave, C., Liuni, S., Licciulli, F., and Attimonelli, M. 2000. Update of AMmtDB: a database of multi-aligned metazoa mitochondrial DNA sequences. Nucleic Acids Res. 1: 153-154.

Lemkin, P.F., Chipperfield, M., Merril, C., Zullo, S. 1996. A World Wide Web (WWW) server database engine for an organelle database, MitoDat. Electrophoresis. 3: 566-572.

Pesole, G., Gissi, C., Catalano, D., Grillo, G., Licciulli, F., Liuni, S., Attimonelli, M., and Saccone, C. 2000. MitoNuc and MitoAln: two related databases of nuclear genes coding for mitochondrial proteins. Nucleic Acids Res. 1: 163-165.

Shimko, N., Liu, L., Lang, B.F., and Burger, G. 2001. GOBASE: the organelle genome database. Nucleic Acids Res. 29: 128-132.

Virus

Albà, M.M., Le, D., Pearl, F.M.G., Sheperd, A.J., Martin, N., Orengo, C.A., and Kellam, P. 2001.VIDA: A virus databsae system for the organisation of the virus genome open reading frames. Nucleic Acid Res. 29: 133-136.

Gashen, B., Kuiken, B., and Foley, B. 2001. Retrieval and on-the-fly alignment of sequence fragments from the HIV database. Bioinformatics. 17: 415-418.

Part III

Online Analysis Tools

Chapter 10

Genomics Tools

Contents

From: *Genomes and Databases on the Internet: A Practical Guide to Functions and Applications*
ISBN 1-898486-31-X © 2002 Horizon Scientific Press, Wymondham, UK.

Abstract

Genomics tools are computational methods that augment experimental genomic analysis. Finding and aligning nucleotide sequences, creating phylogenetic trees, identifying genes from genomic sequence, designing primers, and analyzing microarray data are areas of genomic analysis that are supported by software found on the web. There are multiple tools for each task and variation in their capabilities. This chapter breaks down the differences between tools, explains some of the differences, and lets you know where to find them. The analysis includes Entrez, SRS, PHYLIP, Primer3, CLUSTALW, and more.

Introduction

Online genomics tools are software or applications from the web that can be used to analyze nucleotide data. Nearly all nucleotide data is sequence, with the exception of minimal structure data. This chapter focuses on tools for nucleotide sequence analysis that align batches of nucleotide sequences, create phylogenetic trees from aligned sequences, predict the location of genes from genomic DNA, and predict primers for PCR (Polymerase Chain Reaction). Tools closely associated with genomics tasks are also described here: database query tools, database applications (AceDB), and gene expression (microarray) tools.

The application of these tools tends to overlap with nucleotide and protein sequence analysis. Tools for similarity searches (BLAST and FASTA), alignment (CLUSTALW), and phylogeny (Phylip) can be applied to both nucleotide and protein sequence data. Protein analysis is different because protein research incorporates substantially more structure data than nucleotide research. Proteomics tools include an array of both structure analysis tools and sequence analysis tools. Chapter 11, *Proteomics Tools*, describes the various tools available for protein sequence and structure analysis.

There are alternatives for the genomic tools that can be found in the commercial sector, whose offerings this book does not review, but because of their popularity with the scientific community we mention three of them here. GCG is a suite of computational tools that cover many facets of genomic research, PAUP offers phylogenetic analysis tools very similar to those of Phylip, and Oligo is software for primer design.

The chapter includes the analysis of various applications from the following computational fields:

- Database Query
- AceDB
- Similarity Search
- Multiple Sequence Alignment
- Phylogenetics
- Gene Identification
- Primers
- Microarray Analysis

Database Query

All databases have tools to assist users in finding data particular to their interests. For those whose interest lies beyond data from a single database, query tools have been developed that mine data from multiple sources. Entrez, SRS, and DBGET are text-based tools that assist researchers in extracting data from multiple sources by finding and presenting results through a single web interface, thus saving you valuable time.

Different types of data from many databases are integrated with these query tools, such as nucleotide sequence, protein sequence, structure and classification. These databases also facilitate multiple databases searches focused on one data type. For example, these tools can search both PIR-PSD and SWISS-PROT protein databases, to make sure you have accessed all data in the public domain pertaining to protein sequences.

Entrez

Entrez, from the National Center of Biotechnology Information (NCBI) in the United States is the primary query application for numerous databases. One of the unique features of Entrez is that it only queries data sources offered at NCBI, though with data collaborations those sources are fairly extensive.

NCBI data accessible through Entrez includes:

- PubMed: The biomedical literature (PubMed)
- Nucleotide sequence database (Genbank)
- Protein sequence database (composite of SWISS-PROT, PIR-PSD, PRF, etc.)
- Protein Structure: Three-dimensional macromolecular structures (MMDB)
- Genome: Completely sequenced genomes and those in progress
- PopSet: Population study data sets (sequence alignments)
- Taxonomy: Organisms in GenBank
- OMIM: Online Mendelian Inheritance in Man

As mentioned above, NCBI offers a variety of data sources. Entrez takes advantage of its tight link to NCBI data by having more control of the results of its queries. A search in Entrez does not simply extract data from each source and present it individually, as if it had performed ten separate searches and compiled the results on one long page. Instead the search results are interwoven into a fabric of cross-linked related data. The integration of the data gives you concise and focused results.

SRS

SRS is the Sequence Retrieval System from the European Bioinformatics Institute (EBI). While Entrez integrates databases of the same type and a query is performed on the composite database, SRS leaves databases as individual entities. For example, in Entrez a search of protein sequences is done on NCBI's *Proteins* database, which is a compilation, and you cannot perform a search on PIR-PSD or SWISS-PROT individually. SRS lets you choose which databases you want to search. The databases are indexed making the selection of databases to query quite easy. The results of a search reflect this—you receive a list of entries from each of the databases searched. SRS offers more databases than the other query tools and it is also available as downloadable software.

Here are some of the database types accessible through SRS:

- Literature (i.e., Medline)
- Sequence (various databases for nucleotides and proteins)
- Classification databases (i.e., InterPro)
- Taxonomy
- Transcription Factors
- Protein Structure (i.e., PDB)
- Genomes
- Mapping
- Mutations
- Metabolic Pathways

DBGET (and LinkDB)

The DBGET/LinkDB system is offered by the Japan's GenomeNet service. It is similar to SRS because it searches individual databases and presents the results from those databases in a list. DBGET is the query tool that performs the search. LinkDB is used in conjunction with DBGET by providing links to further information on the individual entries of a query result.

DBGET/LinkDB is from the same group responsible for the KEGG database of pathways and genomes. KEGG is an excellent resource whose data is fully integrated with the DBGET/LinkDB query system. KEGG is fully reviewed in Chapter 7.

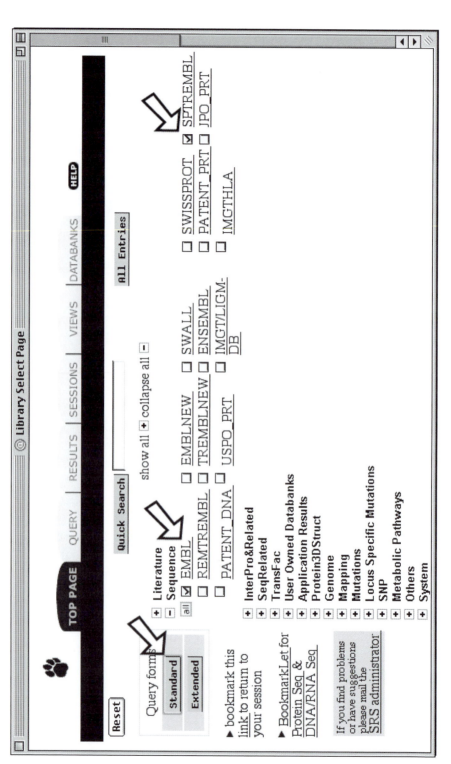

Figure 10.1. Database selection from *Top Page* at the Sequence Retrieval System, SRS. See Chapter 10 *Query Example: SRS* for further information.

DBGET currently supports the following database types:

- Nucleic acid sequences (GenBank, EMBL)
- Protein sequences (SWISS-PROT, PIR, PRF, PDBSTR)
- 3D structures (PDB)
- Sequence motifs (PROSITE, EPD, TRANSFAC)
- Enzyme reactions (LIGAND)
- Metabolic pathways (KEGG)
- Amino acid mutations
- Amino acid indices
- Genetic diseases (OMIM)
- Literature (LITDB, Medline)
- Gene catalogs (KEGG)

Query Example: SRS

To get an idea of how SRS works, you can try the following query example:

From the SRS home page, select *Start*. This takes you to the set of pages dedicated to queries and results. From the first page you see, *TOP PAGE* (Figure 10.1), open the list of databases for *Sequence* and check each box corresponding to the sequence database of choice. Select two primary databases, one representing nucleotide sequences (EMBL) and the other representing protein sequences (*SPTREMBL* for SWISS-PROT + TrEMBL). These databases would otherwise have to be searched separately, but SRS allows both to be searched in one effort. Then select the *Standard* query form that takes you to the page named *QUERY* (Figure 10.2). You can see *EMBL* and *SPTREMBL* are the databases the search will include. From here, select the type of search, *organism* and *keywords*, and enter in *Arabidopsis* and *protein kinase* in their respective fields. There are search parameters that may be adjusted, but you can use the default values and simply click *Submit Query*. The results are a list of links to the individual files at the corresponding databases. So there is a list of EMBL entries that matches the query and another list from SPTREMBL. The results are seen on the *RESULTS* page.

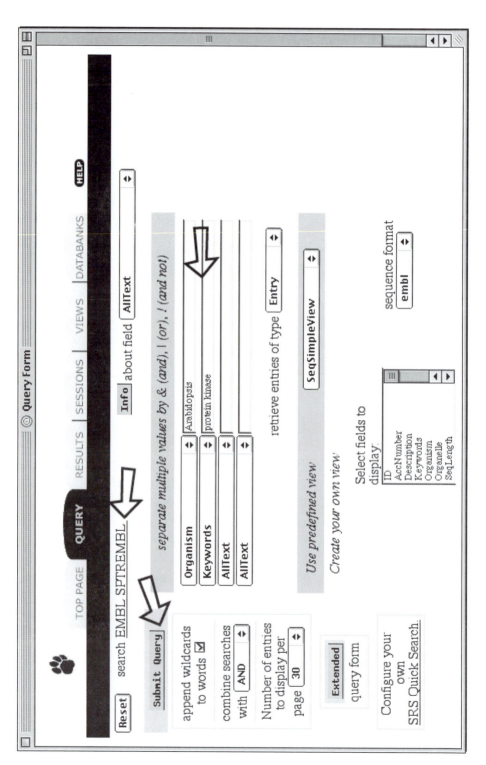

Figure 10.2. Standard query from *Query* page at the Sequence Retrieval System, SRS. See Chapter 10 *Query Example: SRS* for further information.

The example provided is extremely simple, but it gives a clear indication as to how a query tool can assist you in searching multiple databases. Tutorials for how to use and develop the SRS system are available at the SRS site.

AceDB

Databases are indispensable tools for genomic research. The AceDB (A *Caenorhabditis elegans* database) is a freely available genomic database used by genome centers around the world for numerous projects. Originally written for the *C. elegans* genome project, AceDB has become an independent database software package. For example, the plant genome databases at Cornell University utilize AceDB software almost exclusively. AceDB runs on a multitude of computer operating systems including Windows, Macintosh and various types of Unix. Because it is free and readily available, it has become a favorite of data archivists for all types of information from genomes to nomenclature. AceDB can be used locally, from the machine that it is running on, or remotely, via a client-server relationship. One of the features responsible for increasing the usage of AceDB is the development of a web interface to go with its data serving capabilities. This component allows users to query online AceDB servers remotely via the Internet and comes in a variety of user interface configurations.

For downloads and more information about AceDB, visit www.acedb.org.

Similarity Search

Similarity search tools take a nucleotide or protein sequence, compare it against sequences in a database, and extract the most similar entries. In a way these tools are subset of the "Database Query" tools, but because of their ubiquity—they are found in nearly every database— they deserve special mention.

Similarity searches are done most often in an attempt to identify an unknown sequence. BLAST (Basic Local Alignment Search Tool) and FASTA use local alignments, versus global or whole sequence alignments, in comparing sequences. Local alignments use a use more sensitive and faster search algorithm than global alignments but can make wrong assumptions about distantly related data. Global alignments are not as common on the web and take longer to yield results. Using both BLAST and FASTA is recommended for resolving unclear results. There is plenty of documentation available on the web that details the differences in the various alignment methods.

BLAST and FASTA are simple to use. Copy a sequence from a file stored on your personal computer and paste it into the empty box dedicated for a query sequence. Many times you will have to pick the type of BLAST or FASTA you are looking to perform, either a nucleotide sequence search, a protein sequence search, a nucleotide translation then a search, etc. After deciding on the type of search you are interested in, click the "submit" button and wait for the results. Generally, within a minute or so, a list of the most similar sequences will appear, which are linked to the data file describing them in full.

- Common BLAST uses (name):
- Nucleotide sequence search (blastn)
- Protein sequence search (blastp, PSI-BLAST, PHI-BLAST)
- Translated BLAST search: DNA↔ Protein, then a search (blastx, tblastn, tblastx)
- Conserved Domain Database search (RPS-BLAST)
- Aligning two sequences (blast)
- Common FASTA uses (name):
- Nucleotide/Protein seaquence (fasta, ssearch)
- Translated FASTA searches (fastax, fastay)
- Aligning two sequences (align, lalign)

For information regarding the basics of BLAST, see http://www.ncbi.nlm.nih.gov/BLAST/. For FASTA, see http://www.ebi.ac.uk/fasta3/.

Multiple Sequence Alignment

These tools align two or more sequences against each other (whereas BLAST and FASTA only align a total of two sequences against each other). Sequence alignment is most commonly done with proteins in an attempt to find conserved domains leading to the classification of proteins, but there are applications of sequence alignment for nucleotide sequences. For example, sequences can be aligned for molecular evolutionary studies as a preliminary step in the building of a phylogenetic tree. Phylogenetic tools, discussed below, often require sequences to be aligned before the creation of the tree. These tools will not only perform an alignment, but will also generate results according to the format of choice, which is often the format the phylogenetic tool requires. Two tools for nucleotide sequence alignment are DCA and CLUSTALW. Both perform protein and nucleotide sequence alignments while almost every other multiple sequence alignment tool is dedicated to protein sequences only. Web options for protein sequence alignment can be found in Chapter 11, Proteomics Tools.

DCA, Divide-and-Conquer Multiple Sequence Alignment, and CLUSTALW can be used through the web or can be downloaded to a local server. DCA, for example, can only process an alignment that uses less than 500 Mb of its memory. Any use of DCA that requires more memory (larger tasks) must be done on a local server.

There are many parameters that can be adjusted before submitting a batch of sequences to be aligned. To perform a nucleotide sequence alignment through DCA, "DNA/RNA" must be selected from a pull-down menu under the category of predefined matrices (or computational algorithms). There are algorithms specific to proteins (Blosum, PAM, Gonnet) and, in this case, one specific to nucleotide sequences. The default matrix for an alignment is for protein sequences. The various parameters that affect the stringency of the alignment and the type of output a user receives from a submission are best explained by documentation at these sites. CLUSTALW does a good job of identifying each parameter it offers.

Table 10.1. Phylogenetic Analysis Tools

#	Site Name	URL
1.	Phylogenetic Analysis Computer Programs	http://phylogeny.arizona.edu/tree/programs/programs.html
2.	Phylogeny Programs	http://evolution.genetics.washington.edu/phylip/software.html
3.	Phylogenetics Resources	www.ucmp.berkeley.edu/subway/phylogen.html
4.	Phylip	http://evolution.genetics.washington.edu/phylip.html
5.	PAUP	http://paup.csit.fsu.edu/index.html

CLUSTALW offers output files in formats acceptable by phylogenetic tools such as Phylip.

DCA is found on the web at http://bibiserv.TechFak.Uni-Bielefeld.DE/dca/ and CLUSTALW at http://www.ebi.ac.uk/clustalw/.

Phylogenetics

Phylogenetics looks at the evolutionary relatedness of organisms, genes, proteins, etc. In this case we are looking at the phylogenetics of nucleotide sequences. The result of a phylogenetic study can be a tree. A phylogenetic tree such as those developed by Phylip, the most common web accessible tool for this analysis, graphically displays evolutionary relatedness based on sequence similarity. Phylogenetics, like multiple sequence alignment, is quite common to protein sequences as well. Phylip has options for creating trees for both nucleotide sequences and protein sequences.

Phylogenetic tools develop trees from an input of aligned sequences. There are different algorithms for tree drawing such as distance methods, parsimony, and likelihood methods. Phylip has tree drawing tools for each of these algorithms. The different algorithms, how they

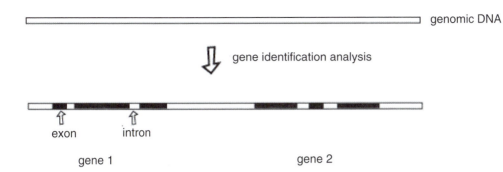

Figure 10.3. Gene identification aims to find genes in genomic sequence. This entails finding splice sites separating introns and exons. Gene identification analysis tools predict the location of introns and exons from genomic DNA.

affect the stringency of the phylogenetic analysis, and the applications specific to them are fully described in Phylip documentation. This information will relate to any phylogenetic analysis tool you might use.

Table 10.1 lists sites for phylogenetic analysis. Tools that are not free are excluded from this table, but one paid tool you may want to consider is PAUP. PAUP is a very common phylogenetic software package containing a variety of tools for tree drawing, and is comparable to Phylip in many ways. It is worth looking into as an alternative (http://paup.csit.fsu.edu/index.html).

Gene Identification

Gene identification refers to finding coding sequences of genes—the sequence segments that ultimately are translated into a protein—in genomic DNA sequence. Sequencing projects of genomic DNA generate sequence that includes incomplete genes, multiple genes, exons and introns within the genes, and various non-coding sequence segments. So how do you find the regions of sequence that code for proteins in this mess?

Table 10.2. Gene Identification Tools

#	Site Name	URL
1.	BCM GeneFinder	http://dot.imgen.bcm.tmc.edu:9331/gene-finder/gf.html
2.	ESTMAP	www.itba.mi.cnr.it/webgene/
3.	GeneBuilder	www.itba.mi.cnr.it/webgene/
4.	GeneParser	http://beagle.colorado.edu/~eesnyder/GeneParser.html
5.	Genie	www.fruitfly.org/seq_tools/genie.html
6.	GenView	www.itba.mi.cnr.it/webgene/
7.	ORFGene	www.itba.mi.cnr.it/webgene/
8.	PROCRUSTES	www-hto.usc.edu/software/procrustes/index.html
9.	Syncod	www.itba.mi.cnr.it/webgene/
10.	MORGAN	www.cs.jhu.edu/labs/compbio/morgan.html
11.	VEIL	www.cs.jhu.edu/labs/compbio/veil.html

Tools for gene identification (see Table 10.2 for a list) take genomic sequence and predict the coding sequence segments that translate into a protein. Methods for this include locating splice sites separating exons and introns and locating exons and introns directly. Once the exons are predicted they need to be put together to form the complete coding sequence. At this point similarity searches are used to identify the gene. If you have protein sequences or ESTs (expressed sequence tags = coding sequence) you can compare genomic DNA to these and extract the sequence segments that match. Protein sequence databases and EST databases (dbEST from NCBI) are used for these purposes. Some of the gene identification tools automatically search databases for homologous sequence, while others will promt you.

Gene identification tools require a query sequence as input, and in some of the tools, the homologous sequences must also be supplied.

Table 10.3. Primer Tools

#	Site Name	URL
1.	Cassandra	www-hto.usc.edu/software/procrustes/cassandra/cass_frm.html
2.	Electronic PCR	www.ncbi.nlm.nih.gov/STS/
3.	GeneFisher	http://bibiserv.TechFak.Uni-Bielefeld.DE/genefisher/
4.	Oligo Calculator	http://mbcf.dfci.harvard.edu/docs/oligocalc.html
5.	Primer3	www-genome.wi.mit.edu/genome_software/other/primer3.html
6.	Primer Generator	www.med.jhu.edu/medcenter/primer/primer.cgi

There are also a variety of parameters that can be adjusted to affect the stringency of the results.

Primers

Primers are the selective agent in a Polymerase Chain Reaction (PCR). They are short segments of single-stranded DNA that can be specific to a unique strand of DNA, for instance a gene of interest. Primer prediction tools can choose the best primers in a given sequence. Other primer tools, Cassandra and GenePrimer, take a genomic sequence and generate primer sequences for presumed exon regions, taking care of both gene prediction and primer prediction at once. The Oligo Calculator characterizes a predicted primer sequence. It determines its melting point (Tm), % GC content, and other physical properties that are useful in setting up a PCR. Web addresses for these tools and others are found in Table 10.3.

Primer prediction software can be used in conjunction with gene prediction tools. You can submit a predicted coding sequence—sequence determined by a gene identification tool—get a set of primers, run a PCR, and sequence the strand to see if the gene prediction was correct.

Table 10.4. Microarray Analysis Tools

#	Site Name	URL
1.	Microarray Data Analysis	http://www.stat.berkeley.edu/users/terry/zarray/Html/index.html
2.	J-Express	http://www.ncgr.org/research/genex/
3.	EPCLUST	http://ep.ebi.ac.uk
4.	Expression Browser	http://www.sanger.ac.uk/Users/mrp/java/ExpressionBrowser
5.	Cluster/ Treeview	http://rana.lbl.gov
6.	Gene Cluster	http://waldo.wi.mit.edu/MPR

Microarray Analysis

Microarrays present molecular biology with a way to perform high-throughput gene expression studies that assist in the understanding of a gene's function.

Microarrays generate so much data that it is impossible to analyze it manually. For example, one array experiment may represent 9000 genes generating 9000 different values of expression dat. Expression is quantifiable allowing array data to be analyzed by statistical measures—an ideal system to apply computer capability.

Typical microarray analysis consists of normalization, filtering, clustering, and biological interpretation. Current tools mostly focus on clustering (the comparison of data across experiments), but they also have capability in the other stages of analysis. Normalization is concerned with variability control within an experiment and filtering with the organization of the data. Biological interpretation of the expression data is primarily a manual process that can be aided by importing information about genes of interest from various databases. Microarray tools assist in this effort with cross-references to databases.

Cluster/Treeview, GeneCluster, EPCLUST, J-Express, and Expression Browser are the major academic tools available over the web. Their URLs are listed in Table 10.4.

Further Reading

Altschul, S.F., Gish, W., Miller, W., Myers, E.W., and Lipman, D.J. 1990. Basic local alignment search tool. J. Mol. Biol. 215:403-410.

Gelfand M.S. 1995. Prediction of function in DNA sequence analysis. J. Comput. Biol. 2: 87-117.

Gelfand, M.S., Mironov, A.A., and Pevzner, P.A. 1996. Gene recognition via spliced sequence alignment. Proc. Natl. Acad. Sci. 93: 9061-9066.

Goodman, N. 2001. The IT Guy: Sampling the Menu of Microarray Software. Genome Technology. March 2001: 40-50.

Pearson, W.R. 1990. Rapid and Sensitive Sequence Comparison with FASTP and FASTA. Methods in Enzymology. 183:63- 98.

Retief, J.D. 2000. Phylogenetic analysis using PHYLIP. Methods Mol Biol. 132: 243-258.

Salzberg, S., Delcher, A., Fasman, K.and Henderson, J. 1998. A Decision Tree System for Finding Genes in DNA. Journal of Computational Biology. 5: 667-680.

Snyder, E. E., and Stormo, G. D. 1995. Identification of Coding Regions in Genomic DNA. J. Mol. Biol. 248: 1-18.

Stoye, J. 1998. Multiple Sequence Alignment with the Divide-and-Conquer Method. Gene. 211(2), GC45-GC56.

Thompson J.D., Higgins D.G., and Gibson T.J. 1994. CLUSTAL W: improving the sensitivity of progressive multiple sequence alignment through sequence weighting, position-specific gap penalties and weight matrix choice. Nucleic Acids Res. 22: 4673-4680.

Walsh, S., Anderson, M., and Cartinhour, S.W. 1989. ACEDB: a database for genome information. Methods Biochem Anal. 39: 299-318.

Chapter 11

Proteomics Tools

Contents

Abstract
Introduction
Protein Analysis Tools at ExPASy
Other Protein Analysis Tools

From: *Genomes and Databases on the Internet: A Practical Guide to Functions and Applications*
ISBN 1-898486-31-X © 2002 Horizon Scientific Press, Wymondham, UK.

3-D Structure Alignment
3-D Structure Annotation
3-D Structure Viewers
Further Reading

Abstract

With the discovery of more genes, faster methods to characterize them are needed. Proteomics tools, computer programs that assist molecular biological research, expedite the process by generating data with minimal experimentation. The tools described in this chapter include more than 100 related to protein sequences and various 3-D structure applications. Applications of the tools vary from translating a determining the amino acid composition of a protein sequence to generating a 3-D structure from a sequence.

Introduction

There are more tools to analyze proteins than any other molecule: tools for identification, characterization, structure prediction, and alignment. Tools are computer applications that assist a researcher in an analysis. The tools found in this chapter are those that are made available for free over the web. They are useful in that they take limited experimental data, such as a protein sequence, and generate further data on a protein, without having to perform unnecessary experimentation.

There are many bioinformatists skilled in designing tools for protein analysis and there are continual improvements in computer capability. Because of this, there are many different tools spread throughout multiple websites.

Most databases also support analysis tools for proteins. Protein sequence, structure, and classification databases offer tools to search for and analyze their data. For instance, a common tool among them is BLAST, which searches databases for similar sequences.

As you can see, there are many independent sources of proteomics tools. There happen to be a number of servers with substantial collections of tools from the independent sources. Of those, ExPASy, Expert Protein Analysis System, has the most comprehensive set. ExPASy was chosen as the server to review for a number of other reasons. It is the home of many important protein databases including SWISS-PROT, PROSITE, SWISS 2D-PAGE, and SWISS 3D. The team that manages ExPASy has developed a number of the tools listed on its site. It continually updates its site so that all the tools it lists are active, and any recently created tools are added. It also provides mirror sites all over the world ranging from Australia, North America, Europe (ExPASy is based in Switzerland), and Asia.

ExPASy http://www.expasy.org

ExPASy has many facets to it, but this chapter only covers its "Proteomics Tools" list. The databases it offers as well as its exceptional portal to all molecular biology links (Amos' Links) are reviewed in other sections of the book.

The other servers with tool collections, those similar to ExPASy, are described at the end of the chapter. Also, though ExPASy does have a very comprehensive set of tools, it does not have some of the significant 3-D structure tools, and these are described at the end of the chapter.

This outline is in response to the complicated nature of the organization of the chapter (see Contents, page 1 of this chapter).

Protein Analysis Tools at ExPASy

It should be noted that the number of tools in the following sections can change and has changed since the review process began. This is a good sign because it shows ExPASy is continuously updating its list. To avoid describing a tool that has since been removed from ExPASy or whose name has changed, the following focus on the general features of the tools with specific mention to better tools.

ExPASy: Protein Identification and Characterization

ExPASY has six groups of tools that deal with the identification and characterization of proteins. Identification involves finding a similar protein in a database dependent on the characteristics of your protein. These characteristics range from a complete amino acid sequence to peptide fragments formed by unspecific cleavage to the molecular weight of the protein. The characterization of a protein involves generating more information of your data. This includes information on its chemical characteristics (i.e., isoelectric point), physical characteristics (i.e., molecular weight), and post-translational modifications. This process often describes a protein from its amino acid sequence. Further analysis may include sending results of this data into one of the identification tools.

The six sets of tools from ExPASy oriented around protein identification and characterization are:

- Protein Identification and Characterization
- DNA → Protein
- Similarity Search
- Pattern and Profile Searches
- Post-Translational Modification Prediction
- Primary Structure Analysis

Protein Identification and Characterization

This section contains eleven tools developed by ExPASy as well as links to four other tools. All fifteen items in Table 11.1 help identify and characterize unknown proteins through the use of their amino acid sequence. Characteristics of a protein include amino acid composition, pI (isoelectric point), MW (molecular weight), and sequence tags.

The first five tools in the Table 11.1 identify proteins based on their characteristics. Table 11.2 below matches a tool with the characteristic(s) it uses for identification. For example, you can input

Table 11.1. Protein Identification and Characterization Tools

#	Tool Name	Tool Description
1.	AACompIdent	identify protein from AA composition
2.	AACompSim	identify proteins with similar AA composition
3.	MultiIdent	identify protein from many variables
4.	TagIdent	identify protein from sequence tag, pI, MW
5.	PeptIdent	identify protein from peptide masses, pI, MW
6.	FindMod	predicts post-translational modifications from MW
7.	GlycoMod	predicts oligosaccharide structures from MW
8.	GlycanMass	calculates mass of oligosaccharide structure
9	FindPept	determines peptides from unspecific cleavage
10.	PeptideMass	calculates peptide masses from sequence
11.	ProeinProspector	identify protein from mass spec data
12.	PROWL	identify protein from mass spec data
13.	PeptideSearch	identify protein from mass spec data
14.	Mascot	identify protein from mass spec data
15.	CombSearch	identify protein from many variables

Notes: tool links found at www.expasy.org/tools/;
AA = amino acid,
pI = isoelectric point,
MW = molecular weight.

Table 11.2. DNA→ Protein Tools

#	Tool Name	Tool Description
1.	Translate	translates nucleotide to protein sequence
2.	Protein Machine	translates nucleotide to protein sequence
3.	MBS Translator	translates nucleotide to protein sequence
4.	Backtranslation	translates protein sequence back to nucleotide sequence
5.	Genewise	nucleotide versus protein sequence comparison
6.	FSED	detects frameshift errors
7.	LabOnWeb	various sequence analyses
8.	List of gene ID sites	extensive list of gene identification software

Notes: tool links found at www.expasy.org/tools/

an amino acid composition and the molecular weight for an unknown protein and the tool searches a public database (i.e., SWISS-PROT), locating proteins with similar amino acid compositions and molecular weights.

Tools six through ten in Table 11.1 define characteristics of the unknown protein. For example, with the FindMod tool your input is the amino acid sequence and MW of the protein, and the results are a prediction of post-translational modifications and a peptide mass fingerprint.

The four tools not developed by ExPASy, items 11-14, utilize mass spectrometry (mass spec) data to identify and characterize a protein. These tools can take mass spec data and search public databases for similar proteins.

The final tool, CombSearch, is an ExPASy developed tool incorporating other identification tools developed by ExPASy in an attempt to integrate selected features of each.

ID Tool   **Input Characteristics**

	Amino Acid Composition	pI	MW of Protein	MW of Peptides	Sequence Tag
AACompIdent	x				
AACompSim	x				
MultiIdent	x	x	x	x	x
TagIdent		x	x		x
PeptIdent				x	

Figure 11.1. A comparison of protein identification tools in terms of the characteristics a tool can use in attempting to identify a protein.

DNA→ Protein

The first half of tools in Table 11.2 focuses on sequence analysis: the sequence translation between DNA and protein. DNA to protein translation is very simple if you have a DNA sequence whose coding sequence has been determined. If you have genomic DNA your sequence must be annotated: remove introns, locate possible frameshift errors, find start codon, etc. The Genewise tool compares a protein sequence to a DNA sequence taking into account introns and frameshift errors. FSED, FrameShift Error Detection, specifically locates frameshift errors. This illustrates the varying capabilities of the translation tools.

The remaining tools in *DNA→Protein* perform the translation and then attempt to identify the protein sequence through a similarity search. These are multi-functional tools. The next section contains tools specific to the latter of the two functions: the comparison of an unknown protein sequence to those in a database via a similarity search.

Table 11.3. Similarity Search Tools

#	Tool Name	Tool Description
1.	BLAST	variety of BLAST links performing similarity searches
2.	Bic	similarity search
3.	DeCypher II	similarity search
4.	Fasta3	similarity search
5.	FDF	similarity search
6.	PropSearch	similarity search from sequence and properties
7.	SAMBA	similarity search
8.	SAWTED	similarity search from sequence and annotation notes
9.	Scanps	similarity search

Notes: tool links found at www.expasy.org/tools/

Similarity Searches

This section picks up where *DNA→Protein* left off: you have a protein sequence and want to identify it. To do this you want to compare it versus all publicly available protein sequences. BLAST, Basic Local Alignment Search Tool, compares proteins by looking for similarity in the sequences. There are many BLAST versions and they are listed here. You can search from a DNA sequence or a protein sequence and specify limitations on the search. These tools search against the primary protein sequence databases (PIR-PSD, SWISS-PROT) so you are comparing your sequence to all of the publicly available data.

The second half of tools in this section is similar in that they perform alignment-based searches, but they add stringency to the search by incorporating extra information. If you have a protein's pI, MW or

Table 11.4. Pattern and Profile Search Tools

#	Tool Name	Tool Description
1.	InterPro Scan	integrated search of multiple databases
2.	ScanProsite	searches for a profile or a pattern
3.	ProfileScan	searches for a profile
4.	Frame-ProfileScan	searches for a profile
5.	Pfam HMM search	searches against the protein family database, Pfam
6.	FPAT	regular expression searches in protein databases
7.	PRATT	generates conserved patterns from unaligned sequences
8.	PPSEARCH	searches for a profile
9.	PROSITE scan	searches for a profile
10.	PATTINPROT	searches for a pattern
11.	SMART	searches for a profile or a pattern
12.	TEIRESIAS	generates conserved patterns from unaligned sequences
13.	Hits	explores the relationship between sequences and domains

Notes: tool links found at www.expasy.org/tools/

textual description, you can input this information along with your sequence and run the similarity search. Table 11.3 notes the tools with extra features.

Pattern and Profile Searches

Pattern and profile searches, listed in Table 11.4, take a protein sequence and determine: (1) whether the protein is part of a family of proteins, or (2) whether it has a conserved sequence segment relating to its structure. The searches operate by comparing the sequence to those in a database. These tools can lead to clues about a proteins' function by confirming its similarity to a group of proteins of known function. Proteins are grouped into families based on similarities in their sequence. For example, a conserved sequence segment may be specific to DNA-binding. If your protein is placed in a family based on the DNA-binding segment, you have narrowed its possible functions down to a DNA-binding activity.

There are also tools that perform an alignment on a batch of unknown sequences in looking for a pattern or conserved sequence among them. InterPro Scan, a query feature of the InterPro protein classification database (see Chapter 5), takes a protein sequence and tries to match it to a protein family. Unlike the other searches, InterPro Scan compares the sequence to multiple classification databases in one effort.

The Hits link offers a protein domain database and a set of tools to explore the relationship between a sequence and its motifs.

Post-Translational Modification Prediction

Proteins often undergo changes after they have been translated. For example, if you determined a protein sequence through a translation of a nucleotide coding sequence, you cannot be sure the protein will ultimately consist of those exact amino acids. This section includes a set of 10 tools, listed in Table 11.5, that predict post-translational modifications. Modification includes cleavage and localization of the protein. Each tool, with an amino acid sequence input and the press of a button, determines where a protein will be cleaved and/or where a protein will be localized within the cell.

PSORT, which looks at localization only, attempts to predict where proteins will be localized anywhere in the cell. The other localization prediction tools determine if a protein will go to chloroplasts or mitochondria only (ChloroP, MITOPROT, and Predotar).

Table 11.5. Post-Translational Modification Prediction Tools

#	Tool Name	Tool Predictions
1.	PSORT	sorting signals and general localization
2.	ChloroP	chloroplast localization
3.	MITOPROT	mitochondrial localization
4.	Predotar	mitochondrial/plastid localization
5.	NetOGlyc	O-glycosylation cleavage sites
6.	big-PI Predictor	GPI modification sites
7.	DGPI	GPI anchor and cleavage sites
8.	NetPhos	Ser, Thr, and Tyr phosphorylation sites
9.	NetPicoRNA	protease cleavage sites
10.	SignalP	signal peptide cleavage sites

Notes: tool links found at www.expasy.org/tools/

The cleavage site tools determine if a protein has a specific type of cleavage site such as the O-glycosylation site, the GPI site, or serine, threonine, and tyrosine phosphorylation sites. There are six tools and they all look for one type of cleavage site only.

Primary Structure Analysis

Primary structure refers to the amino acid sequence of a protein. There are characteristics of proteins that can be predicted by its amino acid sequence alone. These characteristics are then used to help determine a protein's identity and function. Some of the characteristics are amino acid composition, molecular weight, hydrophobic regions, and charge distribution. To analyze a protein, paste in the sequence or enter in an accession number for a sequence from a database, adjust the analysis

Table 11.6. Primary Structure Analysis Tools

#	Tool Name	Tool Description
1.	ProtParam	describes various chemical and physical properties
2.	Compute pI/Mw	generates pI, MW
3.	Protéine	generates pI, MW, AA composition, titration curve
4.	REP	identifies repeats in a sequence
5.	SAPS	describes various properties
6.	Coils	predicts coiled coil regions
7.	Paircoil	predicts coiled coil regions
8.	Multicoil	predicts coiled coil regions
9	PEST	identifies PEST regions
10.	PESTfind	identifies PEST regions
11.	HLA Bind	predicts MHC peptide binding
12.	SYFPEITHI	predicts MHC peptide binding
13.	ProtScale	describes various properties based on scales
14.	Drawhca	plots hydrophobic cluster analysis
15.	Protein Colourer	color codes selected sequence residues
16.	Colorseq	color codes selected sequence residues
17.	HelixWheel / HelixDraw	represents proteins as helices
18.	RandSeq	generates random protein sequences

Notes: tool links found at www.expasy.org/tools/;
AA = amino acid,
pI = isoelectric point,
MW = molecular weight

Table 11.7. Secondary Structure Prediction Tools

#	Tool Name	Tool Prediction
1.	AGADIR	helical content
2.	BCM PSSP	secondary structure prediction and PROSITE search
3.	Prof	secondary structure
4.	GOR I	secondary structure
5.	GOR II	secondary structure
6.	GOR IV	secondary structure
7.	HNN	secondary structure
8.	Jpred	secondary structure from a consensus of prediction methods
9	nnPredict	secondary structure
10.	PredictProtein	secondary structure plus many other related tools
11.	PREDATOR	secondary structure from one or many sequences
12.	PSA	secondary structure and folding classes
13.	PSIpred	secondary structure plus two other structure tools
14.	SOPM	secondary structure
15.	SOPMA	secondary structure with multiple sequences

Notes: tool links found at www.expasy.org/tools/

parameters or leave them at the default settings, then start the analysis and wait for the results. There are 18 links to primary protein structure analysis tools, four of which were derived at ExPASy (see Table 11.6).

Ten of the tools produce chemical and physical characteristics. Five of the ten give textual results with each varying in the characteristic it determines; ProtParam covers multiple characteristics (composition, isoelectric point, molecular weight, etc.) while REP only covers sequence repeats. Similar to most proteomic tools, there is some overlap with each tool. Four of the ten tools report general protein characteristics using color. If you ask these tools to identify hydrophobic regions in a sequence, they will return results where anything in red, for example, is a hydrophobic region.

Seven of the eighteen tools identify specific protein regions: coil-coil (three tools), PEST (2), and MHC (2). These are conserved regions in proteins that help lead a researcher to a description of a protein's function and its identification.

ExPASy: Protein Sequence to Structure

The three types of tools offered through ExPASy that predict protein structural features from sequence are:

- Secondary Structure Prediction
- Tertiary Structure
- Transmembrane Regions Detection

Secondary Structure Prediction

Primary structures are the amino acid sequence of a protein. Secondary structures take into account the interactions among amino acid residues giving the sequence structural life in the form of three secondary structure elements: alpha helix, beta sheet, and coils.

```
> test

  1    VQAIYVPADDLTDPAPATTFAHLDATTVLSRQISELGIYPAVDPLDSTSR    50
           EEEE                            HHHHHHHHH


 51    MLPHILGEEHYNTARGVQKVLQNYKNLQDIIAILGMDELSEDDKLTVARA   100
           EEE          HHHHHHHHHHHHHHHHHHHH          HHHHHHH


101    RKIQRFLSQPFHVAEVFTGAPGKYVDLKESITSFQGVLDGKFDDLPEQSF   150
       HHHHHHH      EEEEEE        EEEEE                   EE


151    YMVGGIEEVIAKAEKISKESAA                              172
       EE      HHHHHHHHHHHHHH
```

Figure 11.2. This is an output of the secondary structure prediction tool *Predator*. The amino acids of the protein sequence are labeled according to their predicted secondary structure state. The H's refer to alpha helices, the E's to beta-sheets, and the underscores to coils.

All of the tools that predict secondary structure (Table 11.7) identify what amino acids from your sequence are involved in these three structural types. The results come in a textual form with the appropriate residues labeled according to their structure. An example of the output is depicted in Figure 11.2.

The use of the tools is quite simple. The input is the protein sequence. Two tools, SOPMA and PREDATOR, allow multiple sequence input.

You can adjust a few parameters before submitting your sequence to be processed by the tool. Each tool has a slightly different algorithm, or computer program, by which secondary structure is determined. Algorithms for all of the tools are explained in documentation at their websites. The tool looks at each amino acid residue and determines whether it is involved in an alpha helix, a beta sheet, or a coil.

Each tool claims a certain percent accuracy for their algorithm. The accuracies range from 65% to 82%. Accuracies are determined by testing the algorithm with a sequence whose structure has been experimentally determined.

Jpred from EMBL is impressive because it is a consensus method of structure prediction. It utilizes 11 different algorithms at once. Surprisingly, EMBL's own method, Jnet, one of the eleven used in Jpred, results in accuracy measurements equivalent to that of Jpred. EMBL developed three of the fifteen tools in this section. PB-IL, Pôle Bio-Informatique Lyonnais, developed six of the fifteen tools listed.

As the structure of a protein becomes clear, so does its function. This is the reason for structure prediction tools.

Tertiary Structure

A tertiary structure consists of the assembly of the secondary structure elements, alpha helices, beta sheets, and coils. Tertiary structure and three-dimensional (3-D) structure generally refer to the same thing: the protein at its functional state, as it would be found in the organism. The items in this section take an amino acid sequence and predict the tertiary structure. See Table 11.8 for the list of tools.

There are other tertiary, or 3-D, structure tools that are not covered in ExPASy's list. These tools, described later in the chapter, perform alignments and annotation on 3-D protein structures found in the Protein Data Bank (PDB). Tools to view 3-D structures are also described later in the chapter.

Tertiary structure prediction and modeling is important, for example, in designing drug discovery experiments that focus on site-directed mutagenesis. They can also be used as an illustration tool.

Modeling is a quick way to analyze a protein sequence by looking at its functioning structure without the time and expense involved in experimentally determining it. This is done through a comparative

Table 11.8. Tertiary Structure Tools

#	Tool Name	Tool Description
1.	SWISS-MODEL	comparative 3D modeling
2.	CPHmodels	comparative 3D modeling
3.	3D-PSSM	fold recognition
4.	SWEET	constructs 3D models of saccharides
5.	Deep View (or Swiss-Pdb Viewer)	3D model display

Notes: tool links found at www.expasy.org/tools/

approach. It uses experimentally elucidated structural data from the PDB as a template and threads your sequence onto it. Template structures are chosen by sequence similarity.

SWISS-MODEL is the item of focus in this review because it is the most thorough of the modeling tools listed in ExPASy. Associated with SWISS-MODEL are the SWISS-MODEL Repository, a database of 3-D models, and Deep View, a 3-D structure viewer.

There are numerous complexities to developing the model and manipulating it through Deep View that are best learned by exploring the site. The site provides excellent materials to guide a novice including background information on protein structure and a tutorial on how to develop and analyze your own model.

First, SWISS-MODEL finds a structural template using your sequence. A template is determined by finding a similar sequence whose structure is known. That structure becomes a template. The sequence similarity search is done through ExPDB and ExNRL-3D, databases local to ExPASy and directly associated to PDB, however, they have manipulated PDB files for better use by SWISS-MODEL.

Next, you provide your sequence, the template(s), and an email address (SWISS-MODEL is free to all users) and within a short time theoretical models are sent back as an email attachment (approximately 250 kilobytes).

Finally, you view and manipulate the 3-D model in Deep View (formerly SWISS-PDB Viewer). Deep View can be downloaded for free from the SWISS-MODEL site and is only two megabytes. Deep View is loaded with various functions in order to view and manipulate a model. A novice can quickly view a file and perform basic visual procedures such as rotation and magnification of the model. Modeling results can be checked for accuracy through the sites listed in Table 11.8.

ExPASy performed a 3D Crunch project in 1998 that aimed to model all protein sequences in SWISS-PROT. Of course, not all structures could be modeled because experimental structure data is limited, resulting in a fewer number of available templates. All theoretical model data is stored in SWISS-MODEL Repository. Protein structure databases are discussed in Chapter 6.

The section *3-D Structure Viewers* below contains information on 3-D viewers for structural data.

Transmembrane Regions Detection

These tools detect transmembrane regions in a protein. Transmembrane proteins span plasma membranes of a cell (i.e., chloroplast, mitochondria, the cell) by way of alpha helices. Determining if your protein has a transmembrane section will provide hints to its function.

The tools, listed in Table 11.9, use a protein sequence as the input and predict (1) the residues involved in transmembrane structural domains, (2) the number of alpha helices, and (3) the orientation of the helices. The orientation refers to whether the flow of the helix starts inside the structure surrounded by the membrane and ends outside of the structure or if it begins outside and ends inside.

Table 11.9.   Transmembrane Regions Detection Tools

#	Tool Name	Tool Prediction
1.	DAS	transmembrane regions in prokaryotes
2.	HMMTOP	transmembrane regions
3.	PredictProtein	transmembrane regions
4.	SOSUI	transmembrane regions
5.	TMAP	transmembrane regions from multiple sequences
6.	TMHMM	transmembrane regions
7.	TMpred	transmembrane regions
8.	TopPred 2	transmembrane regions

Notes: tool links found at www.expasy.org/tools/

Because there is nothing significantly different about any particular tool, they will not be mentioned specifically. All tools are easy to use and give results quickly.

ExPASy: Protein Sequence Alignment

Sequence Alignment

Sequence alignment tools attempt to find similarities between sequences. If you find an amino acid sequence that shows great similarity to another you can presume similar function.

There are two ways to compare the order of amino acids between sequences: global alignment and local alignment. Global alignment refers to matching whole sequences to each other. Local alignment tools match segments of whole sequences. Local alignment is important for protein sequences because conserved short segments

Table 11.10. Sequence Alignment Tools

#	Tool Name	Tool Description

Binary—Alignment of Two Sequences

#	Tool Name	Tool Description
1.	SIM + LALNVIEW	aligns two sequences or segments of one sequence to itself
2.	LALIGN	aligns sub-segments in two sequences
3.	Dotlet	aligns two sequences

Multiple—Alignment of Two or More Sequences

#	Tool Name	Tool Description
4.	CLUSTALW	aligns multiple sequences
5.	ALIGN	aligns two sequences
6.	DIALIGN	aligns multiple sequences based on segments
7.	Match-Box	aligns multiple sequences based on segments
8.	MSA	aligns multiple sequences
9.	Multalin	aligns multiple sequences; color coded results; DNA also
10.	MUSCA	aligns multiple sequences
11.	AMAS	improves quality and presentation of aligned sequences
12.	Bork's Alignment Tools	improves quality and presentation of aligned sequences
13.	CINEMA	improves quality and presentation of aligned sequences
14.	ESPript	improves quality and presentation of aligned sequences
15.	plogo	alignment presented as a sequence logo
16.	GENIO/logo	alignment presented as a sequence logo
17.	WebLogo	alignment presented as a sequence logo

Notes: tool links found at www.expasy.org/tools/

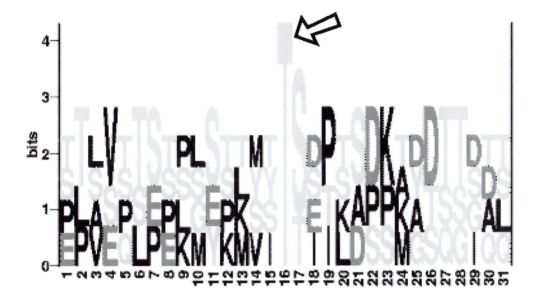

Figure 11.3. A sequence logo generated by *plogo*. The large "T" in the middle means all amino acids at that position in the aligned sequences were Threonines. Sequence logos color code amino acids.

can make up functionally important domains. Proteins are organized into families based on domains.

An important aspect of local alignment is to allow gaps between the matching segments. For example, a functional domain can begin at the 100th amino acid of one protein sequence and the same functional domain can begin at the 105th amino acid in another sequence. If the tool did not allow for gaps amino acids 101-130 of one sequence would not match amino acids 106-135 of the other sequence. Almost all of these tools, listed in Table 11.10, allow for gaps and state it clearly at their web site.

There are other parameters of lesser importance in alignment tools that are explained at the sites. CLUSTALW does a very good job of documentation making it a good site to visit to understand alignment parameters.

ExPASy separates the alignment tools into two sections, binary and multiple. Binary alignment tools look at two sequences and multiple alignment tools look at two or more sequences. For example, ESPript can align as many as 1000 sequences at once.

The "multiple" section of Table 11.10 has three distinct groups of tools. Tool numbers 4 to 10 perform multiple sequence alignments, tools 11-14 take previously aligned sequences and add quality to the alignment and the alignment display, and tools 15-17 represent an alignment as a sequence logo. A sequence logo is useful because its representation of a sequence alignment is visually intuitive and includes all amino acids of each sequence. Other sequence alignments display a consensus sequence, representing only the most frequent amino acid at each position. See Figure 11.3 for an example of a sequence logo.

Sequence alignment often goes hand in hand with phylogentic analysis. Once a batch of sequences is aligned, the sequences can be pasted into tree building software, creating a graphical display of the similarity among sequences. CLUSTALW, primarily an alignment tool, has some tree building capability. Phylogenetic analaysis tools such as CLUSTALW, Phylip, and others are discussed in Chapter 10, Genomics Tools, because tree building can also be done on nucleotide sequences.

Other Protein Analysis Tools

The following review includes descriptions of servers like ExPASy that house a collection of tools, and makes special mention of three types of 3-D structure-based tools—those for alignment, annotation, and viewing—that are not a part of ExPASy's list.

Other Servers with Tools

ExPASy is just one example of a server that compiles protein analysis tools. There are many servers that serve a similar purpose; a few of them are listed in Table 11.11. The ones mentioned below have features setting them apart from ExPASy.

Table 11.11. Servers with Protein Analysis Tools

#	Server Name	Web address
1.	Manchester Bioinformatics	www.bioinf.man.ac.uk/
2.	MIPS	www.mips.biochem.mpg.de/
3.	CMS Molecular Biology Resource	http://restools.sdsc.edu/
4.	SDSC Software and Data Center	www.sdsc.edu/Software/index.html
5.	Tutorial for 3-D structure prediction	www.bmm.icnet.uk/people/rob/CCP11BBS/
6.	Protein Data Bank	www.rcsb.org/pdb/links.html

Manchester Bioinformatics is a server that offers a tremendous tutorial called *BioActivity*. The tutorial is for those who are unsure about how protein analysis tools can be used. The server focuses on protein analysis and lists a series of tools and databases.

BioActivity goes through a step-by-step approach beginning with a DNA sequence. It translates it to a protein sequence, performs a similarity to identify the protein, and characterizes it in terms of its structure and function. It explains the various databases and tools on the web by presenting them at the appropriate step in the analysis. Links to these tools and databases are provided along the way. The tutorial would be extremely beneficial to anyone not familiar with data analysis on the web, with a special focus on proteins, and would like a hands-on approach to the available resources. Because of its organized, step-by-step approach, the tutorial is also recommended for anyone that wants to analyze a molecule using a series of tools in sequence, despite previous knowledge of what the web offers. The tutorial allows you to analyze your own data.

MIPS is another server worth mentioning because it presents a different approach to a server setup. Its PEDANT page lists one tool for each step of protein analysis, 14 in all. Because it only offers one tool per analysis type, it is not covered in detail like ExPASy. For some, this site may be better than ExPASy because it puts a limited number of tools on its server. In this way, you do not have to sort through a number of options.

3-D Protein Structure

The protein analysis tools ExPASy covers focus almost entirely on sequence data. Tools for 3-D protein structure analysis include tools that align and retrieve similar structures, tools that annotate structures, and tools to view 3-D structures.

3-D Structure Alignment

The primary purpose of structural alignment tools is to find similar proteins based on a 3-D structural data. An alignment tool takes a structure, compares it to all proteins in the Protein Data Bank (PDB), the primary source of structure data on the web, and returns proteins that are most similar. Structure alignment tools are useful in identifying an unknown protein.

Structure and sequence alignment tools act on the same principle to identify an unknown protein. The results of a 3-D structural alignment should return more informative results because a 3-D structure is the biologically active state of a protein. The sequence refers only to the amino acid makeup.

It is with this in mind that CE/CL, Dali, and VAST were developed (see Table 11.12 for their URLs). There are databases containing pre-computed protein structural alignments (see Chapter 6), so if you want to find structures similar to a protein already in the PDB you should go to the structure alignment databases first. From those databases, simply enter in the PDB ID for the protein and the pre-computed alignments will be retrieved. If you generated your own structural data and it is not in the public domain, you will need to use the alignment tools.

Table 11.12. 3-D Structure Alignment

#	Tool Name	URL
1.	CE/CL	http://cl.sdsc.edu/
2.	DALI	www.ebi.ac.uk/dali/
3.	VAST	www.ncbi.nlm.nih.gov/Structure/VAST/vastsearch.html

CE/CL, Compound Extension and Coumpound Likeness, are tools from the San Diego Super Computer (SDSC) facility that not only perform a structural similarity search against PDB entries, but also allow you to align two of your own structures to each other. You input your own structure by 3-D coordinates. The structure alignment tools explain formatting requirements of 3-D coordinate data.

A set of alignments often can be viewed superimposed on one another. The tool's site will inform you of the viewers you can use to display the superimposed 3-D structures.

3-D Structure Annotation

3-D structure annotation tools aim to add information about a protein to complement that found in a 3-D structure file. 3-D structure files are usually found in PDB. The list of tools and their websites are found in Table 11.13.

PDBsum and MIA are tools that integrate information from many databases to complement the data in a PDB entry. They take advantage of the database proliferation on the web by designing tools to pull together as much publicly available data for a protein. PDBsum integrates fewer databases because it focuses on protein structure. MIA, the Molecular Information Agent, integrates about 75 different sources of online data in putting together a thorough data package on any molecule of interest, not just proteins.

Table 11.13. 3-D Structure Annotation

#	Tool Name	URL
1.	PDBsum	www.biochem.ucl.ac.uk/bsm/pdbsum/
2.	MIA	http://mia.sdsc.edu/
3.	Promotif	www.biochem.ucl.ac.uk/~gail/promotif/promotif.html
4.	CASTp	http://cast.engr.uic.edu/cast/
5.	GRASS	http://trantor.bioc.columbia.edu/GRASS/surfserv_enter.cgi
6.	STING	http://trantor.bioc.columbia.edu/STING/help/index_intro.html
7.	PRESAGE	http://presage.berkeley.edu/
8.	SAS	www.biochem.ucl.ac.uk/bsm/sas/

Promotif and CASTp generate specific information about a 3-D protein structure. Promotif identifies structural motifs and CASTp analyzes the surface of a 3-D structure for pockets and cavities associated with active or binding sites.

GRASS and STING perform a more general analysis of the entire protein structure.

Presage and SAS apply structural information to protein sequences. SAS, Sequences Annotated by Structure, structurally defines a single or an aligned set of sequences. Its outcome is similar to that of the secondary structure prediction tools because structural information is mapped onto sequence data, residue by residue. For example, SAS would note that amino acids 1-15 of the sequence are involved in an alpha helix. SAS is significantly different from the secondary structure prediction tools in that it takes a protein's 3-D structure and generates data from it. The data is then related to the sequence. SAS also annotates beyond secondary structure by adding information about individual residues as they apply to hydrogen bonding to ligands, active sites, structural domains, etc.

Table 11.14. 3-D Structure Viewers

#	Tool Name	URL
1.	World Index of Molecular Visualization Resources	www.molvisindex.org
2.	Protein Explorer, Rasmol, & Chime	www.umass.edu/microbio/rasmol
3.	AutoDock	www.scripps.edu/pub/olson-web/do
4.	Deep View	www.expasy.ch/spdbv/mainpage.html
5.	DINO	www.biozentrum.unibas.ch/~xray/dino
6.	Molecular Visualization Programs for Unix/Linux	www. pref.etfos.hr/garlic/competition/index.html
7.	Molecular Visualization Software for Garlic	www.pref.etfos.hr/garlic
8.	Linux4Chemistry	www.zeus.polsl.gliwice.pl/~nikodem/linux4chemistry.html
9	Kinemage Home Page	www.kinemage.biochem.duke.edu
10.	Molecular Interactive Collaborative Environment	www.mice.sdsc.edu
11.	MolScript	www.avatar.se/molscript
12.	OpenRasMol Home Page	www.OpenRasMol.org
13.	PDB and MultiGIF	www.dkfz.de/spec/pdb2mgif
14.	Protein Explorer	www.proteinexplorer.org
15.	Qmol Software	www.lancelot.bio.cornell.edu/jason/qmol.html
16.	Triton Software	www.chemi.muni.cz/lbsd/triton.html
17.	Visual Molecular Dynamics	www.ks.uiuc.edu/Research/vmd

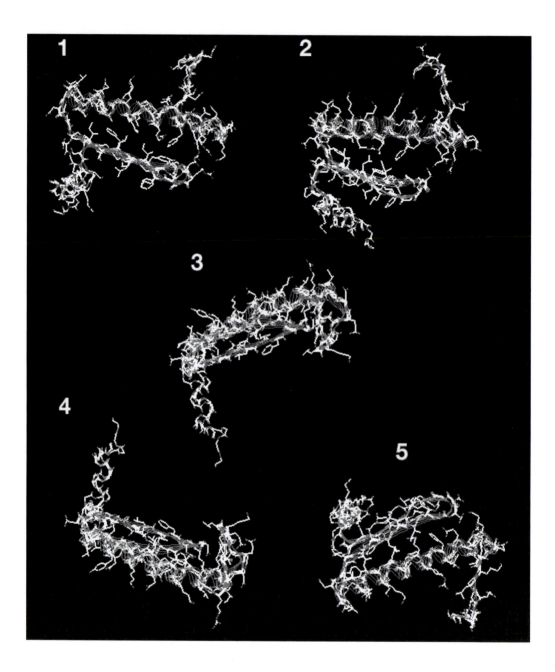

Figure 11.4. This is a series of snapshots from a 3-D structure viewer (*Deep View*). Images 1 to 5 show a protein being rotated. Image rotation is one of many ways users can manipulate 3-D structure images in viewers.

The creators of Presage have a progressive view of what the web can do for science. It is the most interesting of the tools because it provides a means for the scientific community to interact, in and create a virtual structural genomics laboratory for every publicly available protein sequence. Presage, Protein Resource Entailing Structural Annotation of Genomic Entities, provides a means for anyone in the scientific community to structurally annotate a protein sequence. The amount of structural information on each sequence depends on the community's interest in that protein. Researchers can post requests for structural information on proteins of particular interest, possibly leading to collaborations with other researchers. The data from a contributor is not curated. This means you see all data as it was submitted; no data is edited out. Each data submission requires the date the entry was made and the contributor's identity and contact information. The structural annotation can include experimental or theoretical 3-D data. A summary analysis, references, and links to other data resources are also provided. A comprehensive search system allows you to search for a protein sequence and any additional structural information added to it. Searches may include keywords, a contributor's name, or database IDs such as those from SWISS-PROT, Genbank, and PDB.

3-D Structure Viewers

Molecular structure viewing takes 3-D coordinate files, displays them on your monitor, and enables you to alter the view of the molecule. This section is found in the Protein Analysis Tools chapter because these tools are frequently used in conjunction with protein data. Nucleic acids can be viewed with this software so long as they have 3-D coordinate data.

The World Index of Molecular Visualization Resources provides links to tutorials on the use of viewers, sources of 3-D coordinate files (databases), and the software available for molecular visualization and modeling. This and other visualization sites can be found in Table 11.14.

A brief review of molecular structure modeling is found in the description of ExPASy's tool Deep View. See the *Tertiary Structure* section earlier in this chapter for more information.

Of the 3-D structure viewers, Rasmol is most often seen when visiting a website associated with 3-D coordinate data. Rasmol, which was made available to the scientific community free of charge in 1993, is not the best tool available today. Chime is a derivative of Rasmol and can be used in conjunction with the Netscape browser; however Protein Explorer, another derivative, is the easiest to use and the most powerful tool common to the scientific community. It was integrated into the web interface of the Protein Data Bank (PDB) in 1999.

The trend of these tools has been to ease their use by making various visualization functions available in menus instead of typing in commands. The functions include image rotation, magnification,, color coding, etc. The 3-D image of a protein from a viewer is presented in Figure 11.4.

To use these tools, we recommend using tutorials such as those found at the World Index of Molecular Visualization Resources or at the home pages of the tools themselves. The platforms/browsers these various tools run on (Windows/Macintosh/Unix/Linux and Internet Explorer/Netscape) can be confusing. Simply, anything using Chime, including Protein Explorer, requires the use of plug-ins, which are only supported on Nestscape browsers. Some viewers offer downloads and some do not. Generally, using a viewer on your own computer versus over the web is much faster.

Further Reading

Altschul, S.F., Madden, T.L., Schaffer, A.A., Zhang, J., Zhang, Z., Miller, W., and Lipman, DJ. 1997. Gapped BLAST and PSI-BLAST: a new generation of protein database search programs. Nucleic Acids Res. 25: 3389-3402.

Bairoch, A., Bucher, P., and Hofmann, K. 1997. The PROSITE database, its status in 1997. Nucleic Acids Res. 25: 217-221.

Brenner, S.E., Barken, D., Levitt, M. 1999. The PRESAGE database for structural genomics. Nucleic Acids Research. 27: 251-253.

Cuff, J. A., and Barton, G. J. 1999. Evaluation and Improvement of Multiple Sequence Methods for Protein Secondary Structure Prediction. PROTEINS: Structure, Function and Genetics. 34: 508-519.

Cuff, J. A., Clamp, M. E., Siddiqui, A. S., Finlay, M., and Barton, G. J. 1998. Jpred: A Consensus Secondary Structure Prediction Server. Bioinformatics. 14: 892-893.

Fichant, G., and Quentin, Y. 1995. A frameshift error detection algorithm for DNA sequencing projects. Nucleic Acids Res. 23: 2900-2908.

Gorodkin, J., Heyer, J., Brunak, S., and Stormo, G.D. 1997. Displaying the information contents of structural RNA alignments: the structure logos. Comput. Appl. Bioscience. 13: 583-586.

Guex, N., and Peitsch, M.C. 1997. SWISS-MODEL and the Swiss-PdbViewer: An environment for comparative protein modeling. Electrophoresis 18: 2714-2723.

Hansen, J.E., Lund, N., Tolstrup, N., Gooley, A.A., Williams, K.L., and Brunak, S. 1998. NetOglyc: Prediction of mucin type O-glycosylation sites based on sequence context and surface accessibility. Glycoconjugate Journal. 15: 115-130.

Hofmann, K., and Stoffel,W. 1993. TMbase - A database of membrane spanning proteins segments. Biol. Chem. Hoppe-Seyler. 374:166.

Laskowski, R.A. 2001. PDBsum: summaries and analyses of PDB structures. Nucleic Acids Research. 29: 221-222.

Liang, J., Edelsbrunner, H., and Woodward, C. 1998. Anatomy of protein pockets and cavities: Measurement of binding site geometry and implications for ligand design. Protein Science. 7: 1884-1897.

Link, A.J. 1998. 2-D Proteome Analysis Protocols. Humana Press, New Jersey.

Lupas ,A.,Van Dyke, M., and Stock, J. 1991. Predicting coiled coils from protein sequences. Science. 24: 1162-1164.

Milburn, D., Laskowski, R., and Thorton, J. 1998. Protein Engineering. 11: 855-859.

Muñoz, V., and Serrano, L. 1997. Development of the Multiple Sequence Approximation within the Agadir Model of α-Helix Formation. Comparison with Zimm-Bragg and Lifson-Roig Formalisms. Biopolymers. 41: 495-509.

Nakai, K., and Kanehisa, M. 1992. A knowledge base for predicting protein localization sites in eukaryotic cells. Genomics 14: 897-911.

Neshich, G., Togawa, Wellington Vilella and Honig, B. 1998. STING (Sequence To and within Graphics) PDB Viewer. Protein Data Bank Quarterly Newsletter. 84.

Parker, K. C.,Bednarek, M.A., and Coligan. J.E. 1994. Scheme for ranking potential HLA-A2 binding peptides based on independent binding of individual peptide side-chains. J. Immunol. 152:163.

Rechsteiner, M., and Rogers, S.W. 1996. PEST sequences and regulation by proteolysis. Trends Biochem. Sci. 21: 267-271.

Shindyalov, I.N., and Bourne, P.E. 2001. A database and tools for 3-D protein structure comparison and alignment using the Combinatorial Extension (CE) algorithm. Nucleic Acids Research. 29: 228-229.

Thompson J.D., Higgins D.G., and Gibson T.J. 1994. CLUSTAL W: improving the sensitivity of progressive multiple sequence alignment through sequence weighting,position-specific gap penalties and weight matrix choice. Nucleic Acids Research. 22: 4673-4680.

Tusnady, G.E., and Simon, I.1998. Principles Governing Amino Acid Composition of Integral Membrane Proteins: Applications to Topology Prediction. J. Mol. Biol. 283: 489-506.

Part IV

General Resources

Chapter 12

Genome and Database Resources

Contents

Abstract
Introduction
Portals
Literature Search
Suppliers
Genome Publications
    General Genome Publications
        Genome Research
        Genomics
        Nature's Genome Gateway
        GenomeBiology.com
        Genomics Today
        GenomeWeb
    Genome Specific Publications
        Dendrome
        Rice Genome Newsletter
        Human Genome News

From: *Genomes and Databases on the Internet: A Practical Guide to Functions and Applications*
ISBN 1-898486-31-X © 2002 Horizon Scientific Press, Wymondham, UK.

Abstract

Web resources for the molecular biologist are diverse. There are databases of links to web resources; there are databases of literature references; and there are databases of lab supplies. There are also publications dedicated to genome research. The community resources in this chapter are great portals into genome research online.

Introduction

The web offers molecular biologists more than nucleotide, protein, and genome databases and tools that analyze their contents. Online resources also include websites that index these resources (portals). In addition, there are databases that look beyond sequences and structures, genomes and pathways. These are databases of literature references and databases of lab supplies. Lastly, there are online publications dedicated to genomic research.

Portals

Portals come in all shapes and sizes. Their role is to direct you to a defined group of websites. They can be geared toward a subject of interest, such as RNA, or they can attempt to list every database and analysis tool on the web that relate to molecular biology.

Portal websites have been indexed in Table 12.1 according to generalized fields of molecular biology.

Literature Search

Medline is a literature reference database of life science research with a focus on biomedicine. If there is an article in the public domain that relates to molecular biology research, it will be found in Medline. Medline provides the citation and abstract of an article.

Medline is accessed through PubMed (www.ncbi.nlm.nih.gov/), a resource from the National Center for Biotechnology Information (NCBI). PubMed Central, also accessed through PubMed, contains full-text articles of a very small percentage of those journals referenced in Medline. There is an ongoing debate between the Public Library of Science (www.publiclibraryofscience.org) and the journal publishers about the accessibility of full-text versions of articles. The library is backed by over 27,000 signatures from the scientific community who support full-text access. See the Public Library of Science website for more information.

A PubMed search is text-based. Type in anything from title words to author names to journal volume number. The advanced search, *Preview/Index*, has over 20 fields on which to base your query.

Suppliers

Suppliers of materials used in experimental research are on the web. The supply databases can be searched like any other database described in this book. The sites also let you purchase directly from the web.

Most purchasing can be done entirely online with a credit card. However, purchase orders can be used if an account has been set up with the appropriate credit information. Many web sites have special promotions for online ordering, such as free shipping, t-shirts, online currency and many other promotional giveaways. Check the supplier's home page for details because there is usually a link to a promotions page. They also provide MSDS (Material Safety Data Sheets), application protocols, and methods for testing products, among other helpful tips. Table 12.2 lists just a few of the suppliers found online.

Genome Publications

There are two types of genome publications found on the web, those that focus on genomics in general and those that focus on a specific

Table 12.1. Portals

#	Portal	URL
Molecular Biology Centers		
1.	National Center for Biotechnology Information (USA)	www.ncbi.nlm.nih.gov/
2.	European Molecular Biology Laboratory (Europe)	www.embl.org/
3.	National Institute of Genetics (Japan)	www.nig.ac.jp/
4.	ICGEB (International)	www.icgeb.trieste.it/
General Molecular Biology Links		
5.	Amos' Links	www.expasy.ch/alinks.html
6.	Highveld	www.highveld.com
7.	CLUE	http://clue.genome.ad.jp/
8.	Computational Molecular Biology at NIH	http://molbio.info.nih.gov/molbio/
9.	CMS Molecular Biology Resource	http://restools.sdsc.edu/
Bioinformatics		
10.	European Bioinformatics Institute	www2.ebi.ac.uk/
11.	HGMP Resource Centre, UK	www.hgmp.mrc.ac.uk/CCP11/
12.	Bioinformatik.de	www.bioinformatik.de/
13.	International Center for Cooperation in Bioinformatics network	www.iccbnet.org/
14.	Bioperl	www.bioperl.org
15.	International Society of Computational Biology	www.iscb.org/
16.	Bioinformatics.org	http://bioinformatics.org/

Genomes

17. Genomes OnLine Database   http://wit.integratedgenomics.com/GOLD/

18. Genomes + Proteins: www.cad.ornl.gov/~rwd/biology.html
     Tutorials, Databases

19. GenomeNet   www.genome.ad.jp/

20. Amos' Links   www.expasy.ch/alinks.html

Proteins

21. Expert Protein Analysis System   www.expasy.org

22. Protein Information Resource   http://pir.georgetown.edu/

23. Protein Data Bank   www.rcsb.org/pdb/

RNA

24. RNA Society   www.pitt.edu/~rna1/index.html

25. RNA World   www.imb-jena.de/RNA.html

Structural Biology

26. International Union   www.iucr.ac.uk/
     of Crystallography

27. Protein Structure Initiative   www.structuralgenomics.org/

28. Structural Biology   www.sbip.org/
     Industrial Platform

Microarrays

29. Stanford Microarray Resources   http://genome-www4.stanford.edu/MicroArray/SMD/
                                      resources.html

30. Microarray Informatics at EBI   www.ebi.ac.uk/microarray/

31. NCGR Gene Expression   www.ncgr.org/research/genex/other_tools.html
       Resources

Table 12.2. Suppliers

#	Supplier	URL
1.	Qiagen	www.qiagen.com
2.	Operon	www.operon.com
3.	Bio-Rad	www.bio-rad.com
4.	Clonetech	www.clonetech.com
5.	Fisher Scientific	www.fishersci.com
6.	VWR Scientific Products	www.vwrsp.com
7.	Invitrogen	www.invitrogen.com
8.	Sigma-Aldrich	www.sigmaaldrich.com
9.	Agilent	www.agilent.com
10.	New England BioLabs	www.neb.com

genome. The general genome publications include printed journals that have been adapted for web access such as Genome Research, web native news sites such as Genomics Today or genomeweb.com, and web native publications such as GenomeBiology.com. Dendrome and The Rice Genome Newsletter are examples of publications dedicated to specific genome projects. Each of the listed publications has varying levels of access ranging from freely available to subscription based online viewing.

General Genome Publications

Genome Research

Genome Research is a comprehensive journal that prints updated methods, genome research reviews, and new research letters each month. This journal is published by Cold Spring Harbor Laboratory Press and requires a subscription.

Genomics

The Genomics website at Wiley is a resource for the genomics community with free special feature articles and new information each month. The journal requires subscription and features letters by research scientists from genomics projects from around the world.

Nature's Genome Gateway

Nature's Genome Gateway, from the journal *Nature*, is a comprehensive web resource dedicated to genomics. A subscription is required for full text.

GenomeBiology.com

GenomeBiology serves the biological research community as an international forum, both in print and on the web. More recent sections of the web site require subscription.

Genomics Today

Genomics Today is a collection of links to various websites and news with information on genomics. It is published by the Pharmaceutical Research and Manufacturers of America.

GenomeWeb

An online collection of searchable news bulletins focused on various genomic topics such as database design, business mergers and

Table 12.3.   Genome Publications

#	Publication	URL
General Genome Publications		
1.	Genome Research	www.genome.org/
2.	Genomics	www.wiley.co.uk/wileychi/genomics/
3.	Nature's Genome Gateway	www.nature.com/genomics/
4.	GenomeBiology.com	www.genomebiology.com
5.	Genomics Today	http://genomics.phrma.org/today/
6.	GenomeWeb	www.genomeweb.com
Genome Specific Publications		
7.	Dendrome	http://dendrome.ucdavis.edu/Newsletter/
8.	Rice Genome Newsletter	http://rgp.dna.affrc.go.jp/rgp/ricegenomenewslets.html
9.	Human Genome News	www.ornl.gov/hgmis/publicat/hgn/hgn.html

partnerships, recent news, and jobs. GenomeWeb has free online access.

Genome Specific Publications

Dendrome

The Dendrome newsletter is a forest tree genome research update. A publication of the Dendrome project (Institute of Forest Genetics University of California, Davis), it also has free online access.

Rice Genome Newsletter

The Rice Genome Newsletter is an online and print publication by the Rice Genome Research Program (RGP) at the STAFF Institute in Tsukuba, Japan. It is published free online.

Human Genome News

This free web newsletter is intended to inform persons interested in human genome research. Human Genome News is sponsored by the genome program of the U.S. Department of Energy.

Index

3d_ali 84
3Dseq 79

A

ABG 84
Accession Number 21
ACEDB 125
Aedes aegyti 134
AGI (*Arabidopsis* Genome Initiative) 127
AGR (*Arabidopsis* Genome Resource) 129
Allgenes.org 120
Amos' Links 177
ANGIS (Australian National Genomic Information Service) 121
Arabidopsis thaliana 109
ARKdb 122
ASN.1 (Abstract Syntax Notation One) 15-17
Arabidopsis thaliana Genome Annotation Database (see also AtDB)130
AtDB (see also TAIR) 127, 130
Atlas of Nucleic Acid Containing Structures 86
Atlas of Protein Sequence and Structure 31

B

BAC (Bacterial Artificial Chromosome) 125
Bacillus Subtilis 143
Banana 130

BankIt 25
Big Picture Book of Virology 147
BIND (Biomolecular Interaction Network Database) 95
BioActivity 197
BioKnowledge 135
BLAST (Basic Local Alignment Search Tool) 15, 17, 26
Blocks 57, 64-65
BMCD 84
BRENDA 94
BRITE (Biomolecular Relations in Information Transmission and Expression) 95

C

Caenorhabditis elegans 135,136
CAMPASS 83
Cassandra 171
CASTp 200
CATH 83
CDS (Coding Sequence) 16, 22
CE/CL (Compound Extension and Compound Likeness) 198
CGI (Common Gateway Interface) 8
Chime 204
Client-server 6
CLUSTAL 26
CLUSTALW 141, 167, 195
CluSTr 57, 66
CMR (Comprehensive Microbial Resource) 109
COGs (Cluster of Orthologous Groups) 62
Colibri 142,143
CombSearch 180

From: *Genomes and Databases on the Internet: A Practical Guide to Functions and Applications*
ISBN 1-898486-31-X © 2002 Horizon Scientific Press, Wymondham, UK.